Exploring Classroom Assessment in Mathematics

Exploring Classroom Assessment in Mathematics

A Guide for Professional Development

Created by

Deborah Bryant
and
Mark Driscoll

Education Development Center, Inc.
Newton, Massachusetts

NATIONAL COUNCIL OF TEACHERS OF MATHEMATICS
RESTON, VIRGINIA

ASSOCIATION FOR SUPERVISION AND CURRICULUM DEVELOPMENT
ALEXANDRIA, VIRGINIA

Library of Congress Cataloging-in-Publication Data:

Bryant, Deborah.
Exploring classroom assessment in mathematics : a guide for professional development / created by Deborah Bryant and Mark Driscoll.
p. cm.
Includes bibliographical references.
ISBN 0-87353-438-7
1. Mathematics—Study and teaching—Evaluation. I. Driscoll, Mark J. II. Title.
QA11.B854 1998
510′.71—dc21 97-43839
CIP

The publications of the National Council of Teachers of Mathematics present a variety of viewpoints. The views expressed or implied in this publication, unless otherwise noted, should not be interpreted as official positions of the Council.

Printed in the United States of America

IN MEMORIAM

Virginia M. Wolfe

Virginia Wolfe, a Pittsburgh demonstration teacher and a participant in the mathematics-assessment-reform projects this book is based on, died suddenly in 1993. Her legacy is revealed daily in the lives of the teachers and students she touched. Virginia transformed mathematics class into an active, engaging experience. Her wit and belief that all children counted were demonstrated in her unique and creative lessons, which she generously shared with colleagues. Virginia helped us all to realize that we, too, could contribute to the quiet revolution going on in mathematics classrooms. We dedicate this guide to her memory.

Table of Contents

Preface

Recent years have seen increased attention to issues of assessment—in particular, a movement toward means of assessing students by using direct measures of their learning in authentic contexts. Much of the public assessment debate has focused on large-scale performance assessment involving changes in district, state, and national testing systems. A quieter but no less significant revolution has been going on in classrooms. Teachers of mathematics have been working to change both the way they elicit evidence of their students' mathematical thinking and the way they use that evidence to monitor students' progress and guide instructional decision making.

With appropriate time, support, and opportunities for learning, teachers can make dramatic changes in their classroom assessment practices. We have been privileged over the past six years to offer that support for a number of classroom teachers, and then in turn to help them provide professional development and mentoring for their colleagues. This guidebook grows out of that work. It describes how professional development can be designed and facilitated to help teachers change assessment practices.

This publication is based primarily on our work and experiences with the teachers and administrators in the Classroom Assessment in Mathematics Network and the Assessment Communities of Teachers project. We recognize and thank them for their substantial contributions: Shirley Cooper, Karen Jeffries, Anita Jones, Carol Nance, Brian Roget, Sandra Snow, Joyce Stone, and Jo Ethel Thomas of Dayton Public Schools and Ann Farrell of Wright State University in Dayton; Joan Cox, Marieta Harris, Velma Hodges, Louise Holman, Lynn McGoff, Lana Solomon, Alberta Sullivan, and David Wheat of Memphis City Schools; Mary Barbuch, Liz Freeman, LouJane Gatford, Sheran Grant, Nancy Jo Grochowski, Rosann Hollinger, Jeff Kloko, Karen Miller, Janet Murphy, Vince O'Connor, Marie Schimenz, Dorothy Schuller, and Karen Villwock of Milwaukee Public Schools; Ande Bibaud, Diane Briars, Lynn Raith, Debbie Saltrick, Mark Sammartino, Bill Schafer, and Mary Shapiro of Pittsburgh Public Schools; Hans DeGroot, Dianne Fink, Kim Hall, Ellen McDonald, and Vance Mills of San Diego Unified School District; Ruth Ann Duncan of Sweetwater Union High School District, Chula Vista, California; and Lise Dworkin, Sandy Lam, Maria Santos, and Sandy Siegel of San Francisco Unified School District. We also thank Education Development Center (EDC) colleagues Sarah Davis, Sheila Flood, Grace Kelemanik, and Kristi Garrett Ransick for their help in creating this publication.

Deborah Bryant
Mark Driscoll
Education Development Center

Introduction

Why Focus on Classroom Assessment?

Major shifts are under way in the world of assessment that require increased roles and responsibilities for teachers. As attention has turned to the importance of classroom assessment, a significant need has arisen to help teachers develop deeper understandings of the purposes and uses of assessment. Focusing on teachers' assessment methods also meets another need that is central to mathematics education reform: it places students' learning at the center of teachers' professional development. As they become more skilled assessors of students' thinking, teachers can also become more informed critics of their teaching practice in ways that can ultimately enhance students' success.

This guidebook is inspired by the potential of classroom assessment to inform teachers and improve mathematics instruction. The phrase *classroom assessment* refers to the collection of information teachers use to monitor students' learning and to make appropriate adjustments to instruction. This information can be *responses to questions* asked of students, *observable classroom behavior,* or *students' products* from which inferences can be made.

Researchers estimate that teachers spend a third of their professional time involved in assessment-related activities, using assessments almost continuously to guide a wide variety of decisions that directly affect the quality of students' learning experiences (Stiggins 1988, 1993).

Many teachers already use a variety of techniques to find out what their students are learning. In the course of any class period, teachers form judgments about students' understanding and make choices about when and how to proceed. Many teachers cannot, however, clearly describe the techniques and processes they use and the beliefs they bring as they form these judgments. What kinds of evidence are these judgments based on? What criteria are used in interpreting that evidence? What constitutes mathematical understanding? How systematically is evidence collected?

Answering these kinds of questions not only deepens teachers' understanding of assessment but also has important effects on instructional practice and curriculum choices. Teachers

The development of this publication was supported by the Assessment Communities of Teachers (ACT) project, funded by the National Science Foundation under grant no. ESI-9353622. Previous development work was supported by the U.S. Department of Education's Dwight D. Eisonhower National Mathematics and Science Education program under grant no. R168010098. Opinions expressed in this publication are those of the authors and not necessarily those of the National Science Foundation or the U.S. Department of Education.

examine what is important to teach and learn in mathematics. They explore how the questions they ask can limit or expand students' thinking, and they question what kind of instructional opportunities need to be provided, given learning goals and assessment results.

The purpose of this guidebook is to help staff developers design programs that are rooted in teachers' classroom work. It details six professional development experiences, called *teacher investigations,* that help teachers develop an organized way of generating and collecting adequate and relevant data about students' understanding. Reliably and validly interpreting data in turn helps teachers make judgments about classroom practice. In this way, the book is designed to engage teachers in the major decisions and actions that underlie the assessment process.

A Framework for Classroom Assessment

A central theme throughout this guidebook is evidence of students' learning. The National Council of Teachers of Mathematics (NCTM) defines assessment as an ongoing process of gathering evidence and drawing inferences from that information for a variety of purposes. This process is composed of four phases (see fig. 1), each of which highlights important decisions that need to be made in order to gather sound and informative evidence (NCTM 1995).

Fig. 1. The four phases of the assessment process

The *planning,* or *design,* phase focuses on decisions about purpose. For example, a mathematics task can be used as a diagnostic (before), embedded (during), or mastery (final) assessment, and it can be designed to tap into selected reasoning or thinking skills. Different tasks can also demand different kinds of communication skills and assess students at varying levels of conceptual understanding. Decisions about purpose affect the design of the task, the methods used to gather and interpret evidence, and the criteria for judging performance.

The *evidence-gathering* phase is about gathering adequate and relevant information about students' learning. It involves decisions about tasks to be created or selected and

The more information teachers obtain about what students know and think as well as how they learn, the more capacity they have to reform their pedagogy, and the more opportunities they create for student success (Darling-Hammond 1994, p. 22).

about procedures for engaging students in the tasks chosen. In this phase, teachers also establish a method for recording performances and make choices about the way in which results are to be reported.

The *evidence-interpreting* phase focuses on the criteria to be used in examining students' work. These criteria need to distinguish between degrees of quality of evidence. Procedures must be developed for assuring that criteria are applied reliably. In this phase, teachers also use the evidence in students' work to analyze the validity of the task—did it assess what it was supposed to (i.e., did it serve its intended purpose?)? If not, the task needs to be changed to elicit responses that indicate whether students have learned what has been taught.

The *use,* or *action,* phase translates assessment information into subsequent actions. Using data gathered about students' misconceptions or progress toward learning goals, teachers may make decisions about appropriate instructional interventions or revise tasks and assessments. These interventions often trigger another cycle through the assessment process, as more data are gathered to assess the effectiveness of instruction.

The investigations in this guidebook are intended to help teachers come to an understanding of these four phases and their interconnections. For example, one session focuses on the variety of methods for gathering evidence; another concentrates on the inference skills associated with the various methods. Because the phases are integrally linked, staff developers will find that emphasizing these connections often helps teachers come to a deeper understanding of each one.

Reasons for Using This Book

There are three compelling reasons to use this guidebook:

1. The investigations are built on several principles of effective professional development, which should be—

   - ongoing, with themes and efforts that progress from one meeting to the next;
   - responsive to the identified needs of the participants (as articulated by the participants);
   - content rich, with substantial attention to the most current materials and thinking in mathematics;
   - experimental and inquiry based, with a view of teaching that feeds an ongoing process of inquiry, data gathering, and interpretation;
   - learner centered, with approaches to learning based on the best knowledge about how learning occurs;
   - collaborative, driven by the desire for mutual understanding among participants and by the power of collective decision making.

   These characteristics present many challenges to facilitators, who must be prepared to adapt to participants' needs, to orchestrate productive discussions, and to anticipate obstacles and opportunities as teachers engage in the assessment process. This book specifically includes guiding principles for staff developers in creating and facilitating effective sessions.

2. The investigations ground instructional change in students' understanding. Many change initiatives in mathematics education focus on teachers' changing their curriculum and instructional behavior by incorporating innovations such as heteroge-

neous groups, technology, or inquiry learning. The emphasis in these professional development experiences is often on discrete teacher behaviors, not on students' understanding. Challenging teachers' belief systems—about mathematics and about how children effectively learn mathematics—is rarely done.

Students' work is central to this guidebook, and in the authors' experiences, this focus has had significant effects for teachers. Teachers who had previously associated good instruction with teacher-centered interactions—revolving around teachers' questions and a narrow range of brief answers—shifted their focus to more student-centered interactions. Teachers revised their thinking about instructional matters on the basis of their enhanced insights into students' understandings and their enhanced awareness of the wide diversity of thinking patterns students bring to their mathematical thinking. As one teacher summarized her new insights into students' difficulties, "All these years, we thought it was a matter of motivation, and it turns out to be a matter of understanding."

3. The investigations focus on capacity building. The investigations in this guidebook are designed to support the development of teachers' judgment and inferencing skills in addition to providing relevant information about the field of assessment. As teachers modify and create tasks, they have opportunities to sharpen and exercise their judgment about mathematics, the quality of tasks, the evidence of students' thinking, and instructional decisions.

   A collaborative environment supports this capacity building. When teachers are asked to make their inferences explicit to their colleagues, they are challenged to articulate their own mathematical understanding and what they consider to be important in students' work. The role of the facilitator is essential in setting the tone and group process that can support the development of judgment as well as knowledge. This guidebook, by providing broad strategies and detailed suggestions, is designed to prepare facilitators for this role.

REFERENCES

Darling-Hammond, Linda. "Performance-Based Assessment and Educational Equity." *Harvard Educational Review* 64, no. 1 (1994): 5–30.

Stiggins, Richard J. "Make Sure Your Teachers Understand Student Assessment." *Executive Educator* (August 1988): 24–26, 30.

———. "Teacher Training in Assessment: Overcoming the Neglect." In *Teacher Training in Assessment and Measuremente Skills,* edited by Steven Wise, pp. 27–40. Buros-Nebraska Series on Measurement and Testing. Lincoln, Nebr.: Buros Institute, 1993.

Designing and Facilitating the Investigations

Espousing distinctive qualities for professional development is one thing; it is quite another to plan and facilitate meetings and discussions consistent with those qualities. This chapter offers guiding principles for designing and facilitating the six investigations described in this book.

Guiding Principles for Design

We have deliberately chosen throughout this publication to use the word *design* in reference to the staff developer's role. The six investigations we describe provide activity "structures" or "shells." For these investigations to be most successful, we believe staff developers have to take an active role in designing or adapting these activities for their own purposes. This section is intended to give an overview of how a staff developer could use these investigations to design a session with teachers. Included here as well is a "design considerations" template, a planning tool applicable to all the investigations.

Purposefulness

We believe strongly in purposeful professional development, learning experiences designed to address specific learning goals or objectives. The text of the teacher investigations identifies objectives we believe those experiences address. Staff developers should consider why they would choose to use a particular investigation: how does it fit with the group's larger learning objectives? The facilitator should be clear on his or her purposes for using particular investigations, communicate those purposes to participants, and adapt the investigations accordingly.

Connection to the Concerns of Participants

We recommend consistent attention to applying the assessment process in professional development—in particular, by eliciting the felt needs, concerns, and questions of participants and incorporating them as much as possible into ongoing planning. As in the classroom, the challenge is best met by a balanced array of assessment approaches—written evaluations, reflective writing, observations of group work, and so on. For example, as one strategy for gathering data on participants' experiences, at the end of each session, we usually have participants reflect in writing on three questions: What is one thing you liked? What is one thing you would change for next time? What is one thing you are thinking differently about? These reflections have been helpful in eliciting needs, and designing future experiences.

Connection to the Assessment Process

All the investigations contain a graphic like that in figure 1 on page 2, depicting the assessment process. For each investigation, the diagram highlights for the reader which aspects of the assessment process are under consideration. Please note that the assessment process is not linear or necessarily sequential; the highlighted phases in the diagrams do not imply a sequence.

For each phase, teachers should reflect on certain overarching questions as they plan classroom lessons. In order to support teachers' changing their assessment practices, these questions should serve as "framing questions." In figure 2, we have listed framing questions to consider in each phase. These questions can guide your design of the investigations.

Opportunities for Exercising Judgment

Ongoing professional development should create opportunities for participants to exercise, share, and discuss their professional judgment. This is particularly true for assessment-related professional development: the assessment process relies on teachers' skills in interpreting evidence and making valid and reliable inferences on the basis of that evidence. Collegial professional development should allow teachers to sharpen skills in making and testing those inferences and judgments.

Opportunities for Reflection

It is important to provide opportunities for individual and group reflection on learning. The investigations offer suggestions for doing so; in particular, they recommend using reflective writing and large-group reflective discussions. The staff developer could add additional opportunities for reflection throughout the investigation and should offer prompts for reflective writing and discussion that are aligned with the purposes of the investigation and the concerns of the participants.

Choice of Materials

The investigations in this book require a variety of materials; for example, assessment tasks, rubrics, learning goals, and student evidence play a central role. These materials are a powerful component of the professional development experience because they form the basis of participants' common experiences. Discussion and learning arise as teachers present different solutions to the same mathematical task, make different interpretations of a piece of a student's work, or struggle to calibrate their use of a particular rubric on students' common responses. The choice of materials used for the investigations should be made with care and aligned with the purpose of the investigation and the concerns of the participants. In an effort to be as practical and useful as possible, this book suggests materials to be used with each investigation—but these are only suggestions.

Figure 3 is a "design considerations" template for the staff developer to use in planning the investigations in this book. It presents the design considerations discussed above in question form and is intended to be used and reused with each of the investigations.

Planning the Assessment and Gathering Evidence

- What is the purpose of the assessment? Is it a diagnostic (before), embedded (during), or mastery (final) task?
- How should the purpose affect the prompt? (For example, "Tell me everything you know about ..." may be appropriate for a "before" task; observation or impromptu writing assignments might be appropriate for embedded assessment; a problem or project might be more appropriate for a final task.)
- What is the mathematics content? How does the content relate to important, content-unifying themes, such as proportional reasoning?
- What reasoning, strategies, and thinking skills do you expect that the task demands? For example, do you expect to see students apply a particular procedure, or does the task invite a range of problem-solving strategies?
- What communication demands are embedded in the task (e.g., clear and convincing arguments, the use of appropriate mathematical terminology, demonstration with manipulatives, the use of symbolic expressions)?
- Especially if this is a diagnostic (before) task, what are some possible misconceptions that students might bring to the task?
- Especially if this is an embedded (during) task, is it efficient enough to give you a quick but penetrating overview of students' understanding?
- Especially if this a mastery (after) task, what are your standards for mastery?
- What are two or three anticipated responses?

Interpreting Evidence

- What did your students bring to the task? What did you learn about your students' context?
- Did you ask the right question (give the right prompt)? If not, how should the prompt be changed? Should you give it again, in revised form?
- Did you see what you expected in content, in thinking and strategies, and in communication?
- Especially if the task was diagnostic, did you notice any misconceptions?
- Especially if the task was embedded, what misconstructions or missing information do you notice? What is the range of students' approaches?
- Especially if you used a mastery task, how much mastery do you see? Overall, in a class portrait, where (i.e., in which of your standards) are they falling short of mastery?

Using the Results

- If you note persistent misconceptions or misunderstandings, what instructional strategies can be applied most productively?
- If an embedded assessment reveals prevalent misconstructions—students taking instructional cues in inappropriate directions—what interventions are helpful?
- If an embedded assessment indicates missing information or lack of skills, what strategies might be most productive (such as making transparencies of students' responses that demonstrate a particular skill and having students discuss it)?
- In a mastery portrait you have drawn of the class from your analysis of students' work, how far are the students from meeting your standards? Overall, is the difference within your tolerance range? If not, what do you need to stress in instruction?

Fig. 2. Framing questions

Workshop Title: ______________________________________

1. Identify your purpose for using this investigation.

2. Identify needs, concerns, or questions of participants to consider in designing this investigation.

3. What framing questions related to the assessment process will this investigation address?

4. How will this investigation afford opportunities for the collegial exercise of professional judgment?

5. What assessment tasks, student work, rubrics, and other content materials will you use? How do those materials relate to your purpose?

6. How will this investigation provide opportunities for reflection?

7. What are some anticipated participant responses or questions?

8. How will you collect data on participants' experience, questions, and concerns?

Fig. 3. Design considerations

Guiding Principles for Facilitation

We recommend an approach to facilitating learning groups for teachers that is apparently simple in theory but can be challenging to do well in practice. It is based on the following tenets:

- Recognize that your perspective on assessment is simply a perspective and not the only possible perspective.
- Give weight to the perspectives of participants by listening carefully to what they have to say (even, or perhaps especially, when it is challenging to do so).
- Give respect and weight to your own perspective as well by being explicit about what it is and how you arrived at it.
- Treat differences in perspective (between participants, between you and participants) as natural, expected, legitimate, and necessary for the growth and learning of those involved in the experience, including you.

These tenets of facilitation imply several "facilitative" behaviors:

- *Listening.* The facilitator tries to understand the perspectives of participants and to help participants understand one another.
- *Being explicit.* The facilitator tries to be clear about his or her perspective and reasoning.
- *Managing the differences in perspectives.* The facilitator analyzes the interactions with an eye toward differences in perspective, names those differences with the group, and explores those differences by inviting others into the conversation. This fosters a collaborative spirit by stressing the importance of mutual understanding, both between the participants and the facilitator and among the participants.

Managing differences in perspective can be especially difficult when participants make a statement that conflicts fundamentally with the values held by the facilitator. The following are examples of such challenging questions and comments from our experience:

- How do you know that this, too, won't pass, as have previous attempts at reform?
- Why do I need this? After twenty years in the classroom, I know a B when I see one.
- His (her, their) standards for students' responses are not as high as mine.
- One rubric is not enough for my class. I need at least two: one for the faster kids and one for the slower kids.

Challenges such as these can often leave facilitators wondering about what is an appropriate response—honest yet still in the facilitative spirit. Below we look at one of these questions and examine it in light of the facilitative approach recommended above.

An Example

"How do you know that this, too, won't pass, as have previous attempts at reform?" Most teachers and administrators—especially the more experienced ones—have seen efforts to reform mathematics education die slow deaths in the face of systemic resistance. Some were personally invested in the success of those efforts, so their skepticism is especially well earned.

In response, we have tried first to listen actively to the skepticism, to the experiences on which the skepticism is based, and to the mindsets about change that are engendered and then to respect the skepticism as legitimate. We acknowledge that very little about assessment reform is inevitable.

We also are explicit about our perspective: some definite shifts are noticeable, and seemingly irreversible. We appeal to the following summary of shifts in assessment (NCTM 1995, p. 83):

- From "norm referencing" of results to "standards referencing" through growth toward specific performance standards
- From a limited range of sources to gathering evidence from balanced, multiple sources
- From reliance on externally determined sources to information derived by teachers from instruction

We explain to the participants that one of our primary purposes is to collaborate with them in aligning their instruction and classroom assessment with these shifts through a set of classroom investigations. Thus, we try to frame our collaboration with them more broadly than as a preparation for yet another set of mandated changes. Often, skeptical teachers are satisfied with this prospect of more workable classroom practice. They see the time required as worthwhile investment in efforts to keep their practice tuned, no matter what future assessment mandates might be.

It is helpful to be explicit with participants about your role in this nontraditional approach to facilitation. Problems can arise when a facilitator's intended role doesn't match the role expected by the participants. Although we prefer the kind of facilitative role that allows an alliance to develop between facilitator and participants—characterized by asking questions, defining problems, and collaborative problem solving—we have learned that participants often expect another model of facilitation.

REFERENCE

National Council of Teachers of Mathematics. *Assessment Standards for School Mathematics.* Reston, Va.: National Council of Teachers of Mathematics, 1995.

Teacher Investigations

This section describes six teacher investigations designed for workshop settings. The pages that follow describe the investigations in detail; here, we briefly describe their distinguishing characteristics.

Experiencing a Task. Teachers experience assessment from the learner's perspective, working together on a mathematical task and reflecting on the mathematics knowledge gained. They concentrate on the qualities of good assessment that set it apart from routine instructional challenges.

Observing Problem Solving. Teachers experience a specific, nontraditional type of assessment based on a broad range of performances from using mathematical tools to problem solving to communicating to applying thinking skills. Teachers thus have an opportunity to weigh the advantages and the challenges in assessing students' complete work.

Examining Students' Work. Interactions among teachers are a central part of this investigation. They work in groups to weigh the evidence in students' work samples and to make either instructional decisions or scoring decisions.

Developing Tasks. This investigation probes the nature of good assessment questions and challenges. It asks teachers to consider what they know about their students, their teaching contexts, their curricula, and their mathematics values and to align all those elements in tasks that effectively prompt students' performance.

Developing a Rubric. A more advanced investigation than Examining Students' Work, this investigation sharpens the discussion of what is valued by teachers as good work in mathematics. They construct or adapt their own rubrics for a task and discuss in greater depth the options for evaluating good work.

Planning Assessment. This investigation extends the discussion of what is valued to clear expressions of instructional outcomes, based on standards for mastery. It allows teachers to make connections between curriculum and assessment: one option is to guide teachers to alter one of their curriculum units to include alternative embedded and summative assessments.

How to Identify a Sequence of Investigations

Each investigation has a force of its own, and any group of teachers would benefit from taking part in an isolated workshop based on one of the models. There is a cumulative force, however, in judiciously sequencing the investigations. We recommend that users of this document plan workshops as an ongoing effort by a group of teachers.

The sequencing of the investigations is important. The logic that guides the sequencing will depend on the needs and goals of each group of teachers that undertakes them. Determining those needs and goals should be the opening move of the staff developer; a choice of sequence should then follow, subject to alteration as the group develops its own character. Below, we offer a possible context for professional development and suggest a sequence that might fit that scenario.

Possible Sequences

Imagine that a school district has been put on notice by its superintendent that "assessment is going to change" over the next few years. Little other information has been passed on, but it is widely known that the state has been piloting some open, constructed-response tasks in mathematics. Teachers and administrators see change coming, and they want to be prepared.

The district mathematics supervisor decides to convene a group of teacher-leaders from schools across the district for a set of monthly meetings aimed at developing an understanding of alternative assessment. She conceives of the meetings as an ongoing course, complete with readings and homework, including collecting students' work. The course will require a final project as well: by the end of the course in the spring, each participating teacher will take a unit from the current curriculum and alter the unit's final and embedded assessments by incorporating several alternative assessment tasks. Furthermore, they will complement these assessment changes with comparable instructional changes to ensure students' opportunity to learn what is going to be assessed. For example, they might plan for more open-ended questioning or small-group work in their instruction.

Consequently, using the sections of this book as a guide, the supervisor may develop the schematic in figure 4 for the course. The sequence starts with an experience of doing mathematics tasks (Experiencing a Task) and continues with analyzing students' work (Examining Students' Work). The course ends with planning backward from student outcomes to help teachers develop assessments for a curriculum unit (Planning Assessment).

Between the second and final investigations, the supervisor has designed two alternative sequences of investigations. The first focuses primarily on assessments for the purpose of making instructional decisions. The supervisor selects two investigations that reveal information about students' learning: Observing Problem Solving and Developing Tasks. The

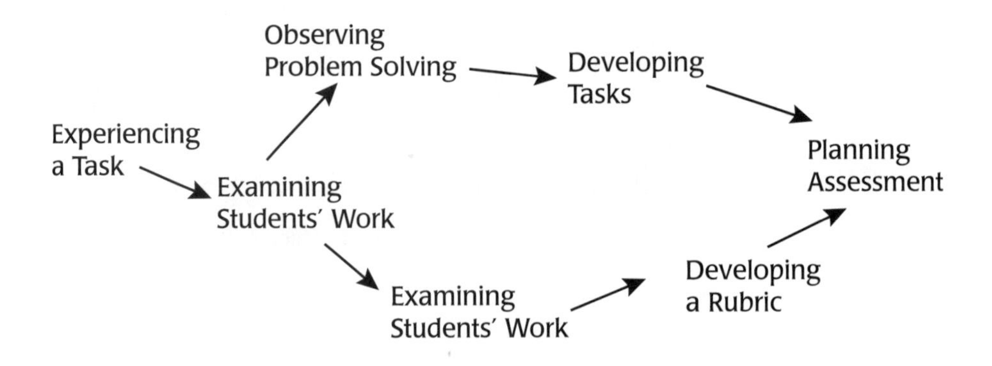

Fig. 4

mathematics supervisor is aware that this group of workshop participants is also concerned about using assessment for purposes of evaluating students' progress and rating their achievement. In the second sequence, therefore, she uses two investigations that focus on grading and scoring: she repeats Examining Students' Work—with an emphasis on evaluative scoring—and incorporates Developing a Rubric.

Another Possible Sequence

A seventh-grade mathematics lead teacher is working with his mathematics department to implement a new curriculum. The features of the new curriculum materials include open-ended questions throughout the text and end-of-the-chapter performance-assessment items. Very few teachers in the department have used these types of items before.

This teacher pulls together the other seventh-grade teachers once every two weeks in the regular department meeting time, and they work on assessment issues related to the new curriculum. They adopt a format that makes extensive use of two of the investigations: Experiencing a Task and Examining Students' Work. Each of their two-hour meetings is split in half: in the first half they look at students' work on an open-ended problem or performance task from the unit they are currently teaching, using the Examining Students' Work investigation. In the second half of the meeting, they use Experiencing a Task to work together on an upcoming problem in the curriculum, doing the mathematics together, reflecting on the important mathematics, and anticipating students' responses. In between meetings, the teachers collect students' work on the task they experienced together for discussion at the next meeting. When appropriate, the lead teacher pulls in other investigations to address topics or concerns as they come up.

Final Tips on Sequencing Investigations

Again, the choice and sequence of investigations is an important decision for the facilitator, and it should depend on the group's purpose and the participants' needs. It may be helpful to remember the following:

- These investigations are flexible and can be altered or adapted to meet multiple purposes.
- These investigations are not designed to be used in any particular order, but that does not imply that order is unimportant.
- These investigations do not compose a complete professional development curriculum but can be used along with other investigations, speakers, and activities.

In particular, using case discussions can be useful learning experiences, especially if the case study contains evidence of students' mathematical understanding or addresses issues in the use of alternative forms of classroom assessment. The Resources section of this publication contains several sources for case materials and case facilitation guides, including books by Carne Barnett (1994), Barbara Miller and Ilene Kantrov (forthcoming a, b), and the Harvard Mathematics Case Development Project (n.d.).

EXPERIENCING A TASK

Overview

In this workshop teachers experience assessment from the learner's perspective, working together on a mathematical task and reflecting on the mathematics knowledge gained. They concentrate on the qualities of good assessment that set it apart from routine instructional challenges.

Experiencing a task as a learner gives teachers firsthand knowledge of the mathematics content, reasoning used, and tools needed to solve the problem. Having worked through the task, teachers also have a clearer sense of the constraints of the task, and they can speak to issues such as students' entry points into the task, collaboration, and attitudes about completing the task.

Objective

- The teachers identify important characteristics of assessment tasks (e.g., purpose, mathematical content, problem-solving strategies used) and the relationship between these characteristics and the ability of the task to elicit new evidence of students' understanding.

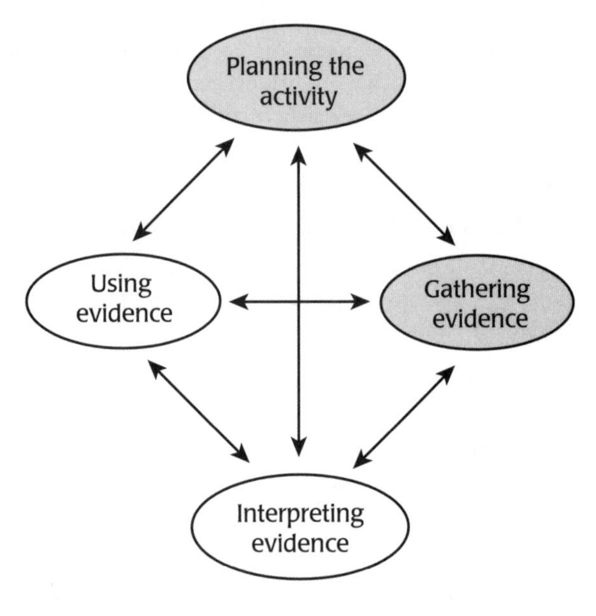

Rationale

As teachers grapple with issues surrounding the use of alternative forms of assessment, such as performance tasks, the shared experience of working through a task can ground discussions in common understanding of the learner's experience. A group of teachers can draw on their collective experience to identify a range of interpretations of, approaches to, or solutions for the task. They also have core pieces of evidence, namely, their experiences with the task and those of their colleagues, from which to make inferences about the appropriateness of the task for their own students. Thus, a primary component of this type of investigation is the structure or framework that focuses reflection and discussion after the mathematics task is completed.

Designing the Investigation

The workshop description on pages 16–17 suggests how this investigation might be organized. The considerations that follow, used along with the design template in figure 3 on page 8, are intended to help you customize this investigation.

1. *Determining the purpose.* What is your purpose for using this investigation? Be clear about identifying your goals and desired outcomes because they drive your selection of tasks and shape after-task reflection and discussion.
   - Do you want to introduce a new form of alternative assessment?
   - Do you wish to illustrate the breadth and depth of students' understanding and reasoning that get communicated during performance assessment?
   - Do you want teachers to reflect on their own understandings in a particular content area?

2. *Finding an appropriate task.* Once you have identified your goals, select a task that highlights them. For example, if you wish to encourage teachers to use performance assessment, select a manageable task that does not rely heavily on hard-to-get materials or assume new instructional strategies that teachers may not be using (such as cooperative learning or writing). Or, if you are focusing on a mathematical "big idea," such as proportional reasoning, select a task that will lead to conversation about that idea. Consider the constraints within which you will work, including time, the availability of materials, grade level, and the teachers' mathematical backgrounds.

   Two sample performance tasks that we have used successfully with middle-grades teachers are shown in figure 5. Note that both these tasks can take a long time because groups can be very detailed about their measurements, drawings, and constructions. We have found it useful to set a time limit, usually of about thirty to forty-five minutes.

Half-Size Me

Work in groups of four. Take measurements of different parts of one person's body, including height, width of shoulders, length of arms, and so on.

On a piece of paper at least as long as half the subject's height, draw a centerline down the length of the paper for symmetry. Be sure you start with the head high enough so that the feet will fit on the paper. Make a drawing of the measured person with all dimensions half the actual size. As you make the drawing, check to see if the dimensions look right. Does the person look too thin or too wide? Are the arms too short? When the drawing is complete, color it and attach your measurements.

This task was developed by Joanie Janis at University of Southern California at Davis and adapted by San Francisco teacher Sandy Siegel.

An Engineering Task

From a given amount of stiff, colored paper, design a closed container with the largest possible volume.

This task was developed by Grant Wiggins of the Center for Learning, Assessment and School Structure.

Fig. 5

3. Choosing discussion topics and questions. The topics and questions you pose are essential to focusing discussion after completing the task. You can target issues related to your identified goals and outcomes. Your goals and the task you select together form the basis for the whole-group discussion and reflection. Which discussion questions in the workshop description are appropriate?

Preparation

- Have teachers work at tables in groups of four. Working in groups allows them to see one another's strategies in approaching the tasks; groups larger than four, however, are more difficult to manage.
- Keep all the materials needed for the task on one common table instead of passing out the materials to each group. This requires the participants to make their own decisions about what tools will be most useful to them as they work.
- Be sure that you have enough copies of the task so that each participant can take home a clean copy.
- Prepare overhead transparencies of tasks.

Materials

Two copies of the tasks for each teacher

Materials needed to complete the task

Blank overhead transparencies and markers for observers' recordings

Other overhead transparencies as needed (of the task, the discussion prompts, etc.)

Time needed

Approximately two hours. This time will vary depending on the task used and the discussion questions selected.

Workshop Description

Getting started

Allow approximately ten minutes. Discuss the goals for the session, and explain the format of the investigation. Gather participants into groups of four. Introduce the task, noting that the task sheets include instructions. Point out the variety of materials that are available for use.

Experiencing and observing the task

Allow approximately forty-five minutes. As you circulate through the room observing each group, consider the following:

- Do not offer instruction other than to address logistical concerns.
- Note what teachers are doing that will guide the coming discussion.

Reflecting on the experience

Allow approximately forty-five minutes. Divide this session into two parts—a presentation of solutions and a reflective discussion.

When the participants have completed the task, have the groups present their solutions briefly (perhaps with one person reporting back for each group) using either an overhead transparency (for An Engineering Task) or flip-chart paper (for Half-Size Me). Have each presenter reflect on the strategy the group used to solve the problem. Depending on the task, you may choose to have the groups record solutions on overhead transparencies.

After each group presents, facilitate a whole-group discussion to reflect on the experience. First, ask participants to reflect in writing, silently and individually. Then, in the group discussion, use prompts such as the following:

- What did it feel like to participate as a learner in an open-ended task?
- What mathematical concepts or skills does this task assess?
- Reflect on the range of mathematics you and your colleagues called on as you performed this task.
- Reflect on the range of strategies and approaches that your group brought to bear on this task.
- Which of these strategies might your students use? Are there any others they might use?
- In what context would this task serve as an appropriate assessment of students' mathematical thinking?
- How should the purpose of the task affect the prompt?
- What could the task reveal about students' understanding?
- What would typical responses from students look like?
- How does students' understanding get communicated in the context of this task?

Summary

Allow approximately twenty minutes. Summarize the issues that were raised during the reflection period and relate the discussion explicitly to your original goals. Discuss possible follow-up options and next steps.

Facilitating the Experience

This teacher investigation seems simple: teachers work through a mathematics performance task and after having completed the task, discuss the experience. This apparent simplicity is deceptive; this investigation requires careful facilitation to ensure both that the teachers focus their attention on their experiences as learners and that the discussion is reflective and substantial.

Keeping the discussion focused on mathematical thinking is one of the biggest challenges for facilitators because it can be difficult for teachers to focus on the underlying *mathematics* of the task. They often want, quite naturally, to discuss instructional issues or the affective impact of the task on students. To counter this tendency, choose discussion prompts that focus first and explicitly on their experience with the task as learners. The workshop description also suggests having participants respond to discussion prompts in writing first in order to support a focused discussion.

During the task, you may want to prompt some groups to consider multiple ways of solving the problem. This can also be a good task extension for those groups that finish early.

OBSERVING PROBLEM SOLVING

Overview

Many teachers observe their students regularly. Typically, those observations are not formally structured, are rarely documented, and tend to focus on ensuring that students are on task and meeting behavioral objectives. Observation data can provide insight into students' mathematical understandings and skills. Such information is useful as teachers make instructional decisions, either for the class as a whole or for an individual student.

In this investigation, the ideas for which were developed by Doug Clarke, Australian Catholic University, and Linda Dager Wilson, University of Delaware, and are used with permission, teachers experience a specific, nontraditional type of assessment based on a broad range of performances from using mathematical tools to problem solving to communicating to displaying thinking skills. They work together on an open-ended question that requires some group discussion. The observers try one or more forms of observation and then report the information they gathered to the group. The participants who solve the problem also reflect on their experiences and on whether the observers' interpretations of their mathematical processes appear valid.

Objectives

- The teachers examine what sources of information they rely on to make judgments about students' understanding.
- The teachers experiment with structured observation techniques to collect data on several dimensions of mathematical thinking and behavior.

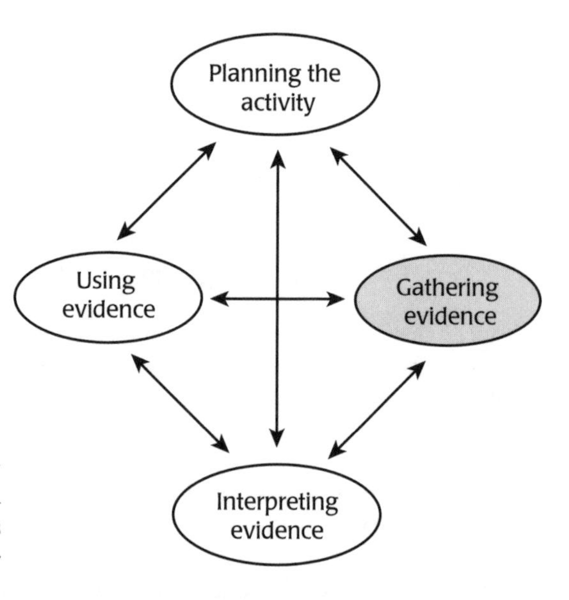

Rationale

There are many ways to gather evidence about students' mathematical understandings: watch what students do, ask questions and listen to their responses, and examine their products. The emphasis in many classrooms is on examining the products of students' work; observing and listening often are not as valued. Observing students as they work on a problem yields rich data for the classroom teacher about what students understand, what they can do, and where they need instruction. Using observation as an assessment to guide instruction is particularly appealing because it does not interrupt the flow of activity in the classroom. As an "embedded assessment," it lets students continue to do important mathematical work while furnishing teachers a lens on students' thinking and communication.

It is not, however, easy to observe students during an activity; in large classrooms with students with diverse needs and various levels of mathematical understanding, observing them in a more formal, organized manner may even seem impossible. Many teachers find it difficult to separate themselves from the classroom activity and just listen. Success with this technique requires practice and opportunities to share results with colleagues.

Designing the Investigation

The workshop description on pages 21–23 is a guideline for how this investigation might be organized. The considerations that follow, used with the design considerations template on page 8, are intended to help you customize this investigation.

1. *Determining the purpose.* What is your purpose in using this investigation? Be clear about identifying your goals and desired outcomes because they drive your selection of tasks and prompts for observers, as well as shape after-task reflection and discussion.
   - Do you wish to illustrate an alternative assessment technique?
   - Do you want to provide an opportunity to discuss the challenges participating teachers encounter when they do formal classroom observations?
   - Do you want to encourage participants to broaden their formal assessment strategies to include more regular use of observation?
   - Do you want to provide specific practice of an observation technique?
2. *Finding an appropriate task.* Once you have identified your goals, select two tasks that highlight them. Each exercise will use one task. The tasks should be open-ended and lend themselves to investigation and discussion. Choosing appropriate problems is essential to the success of observations, both in a workshop setting and in the classroom.

   For the first activity, use a task that requires participants to share the data they have available in order to solve the problem and that requires them to debate the accuracy of their answer. "Fermi problems" lend themselves well to these exercises because a solution to such problems can be approximated by a series of simple, common-sense steps that incorporate data that are generally known or can easily be estimated. (The Mathematics Curricululm and Teaching Program includes lessons that use "Fermi problems." Information on these materials is included in the Resources section.) An example of a Fermi problem follows:

   How many people become teenagers on a typical day in the United States?

   For the second activity, choose a task that requires participants to exhibit particular skills, such as measurement. Whereas the first task requires only discussion and therefore offers a good opportunity to listen to participants' reasoning and measure their communication skills, the second activity should allow for observing particular skills or techniques as well as discussion. One possible question is the following:

> My friend Chris has a balcony on his apartment that overlooks the stadium where concerts and sporting events are held. You can see all the events from his balcony! There is a big concert coming up, and he wants to invite his friends over to watch it. He wants to know how many people will fit on his balcony. It is a large balcony; in fact, it just so happens that it is the same size as this classroom. How many people could Chris fit on his balcony to watch the concert?

3. *Choosing discussion questions.* The questions you ask are important in focusing discussion after completing the task. You can target issues related to your identified goals and outcomes. Your goals and the task you select together form the basis for whole-group discussion and reflection during the investigation. For example, if your goal is to provide participants with an opportunity to share previous experiences with observation techniques and to strategize together about improving those techniques, you may want to emphasize questions about observation techniques in the reflection and large-group discussions. If, however, you are working with a group of teachers who have no experience in observing, you may want to focus discussion questions on planning for the use of such techniques.

Preparation

- Recommend that teachers work at tables in groups of four or five. It is essential to these activities that participants work in groups.
- Be sure that you have enough copies of the Observing Groups and Design Your Own Checklist handouts (see appendix 1 at the end of this investigation) so that each participant can take home a clean copy.
- You may want to identify and photocopy articles or information on observation techniques for participants. Check the Resources section of this document for suggestions.
- Prepare overhead transparencies of the Observing Groups and Design Your Own Checklist handouts.

Materials

Any materials or manipulatives required by the tasks

Photocopies of the tasks you have selected

Photocopies of the Observing Groups sheet

Photocopies of Design Your Own Checklist

Blank transparencies for recording observers' comments

Flip-chart paper

Markers for both flip charts and overhead transparencies

Time needed

Two and a half hours

Workshop Description

Described here are an opening activity and two exercises designed to demonstrate the power of classroom observation as a basis for guiding instructional decisions.

Opening Activity

Allow approximately fifteen minutes. Have the participants work in pairs, each describing one student they know. They ask their partners the following questions:

1. What mathematics does the student know?
2. Which mathematical processes are his strengths?
3. What is her mathematical disposition?

After the pairs discuss the questions, the group facilitator asks the following questions:

- Which of the three questions did you feel most confident in discussing?
- As you were answering the questions, what sources of data were you drawing on?

Frequently teachers will respond that their sources of data include written work, students' responses to questions, and observations of what students do. Often, a substantial amount of data on students comes from observing, although it is frequently not recognized as a strategy for gaining insight into students' thinking.

First Exercise

Getting started

Allow approximately ten minutes. Introduce the task. Choose one person from each table (or roughly a quarter of the participants) to be an observer. Ask the observers to come to the front of the room for further instructions, and have remaining the participants begin the task. Distribute copies of the task to the participants.

Share the observer sheet with the participants acting as observers. This sheet lists the dimensions of mathematical behavior that they are to observe and make notes about while the remaining teachers discuss the problem in groups. Assign each observer one category of mathematical behavior to observe (e.g., "mathematics used" or "problem-solving strategies"). If you have a large group, assign more than one person to each category. Observers are to circulate in the room, listening intently to the discussions but not contributing to the ongoing conversations, even when they hear misleading information or incorrect reasoning.

Experiencing and observing the task (solving the problem)

Allow approximately fifteen minutes. The participants identified as observers circulate, taking notes while the remaining participants discuss the task in groups.

Reflecting on the experience

Allow approximately thirty minutes. Divide this session into two parts—a presentation of solutions and a reflective discussion.

When all the groups have decided on an answer to the problem, each small group shares the solution they found to the problem and their reasoning for solving the problem in the way they chose. After all the groups have presented, have them consider the differences among their answers:

- How do we determine which is the most accurate answer?
- How accurate does the answer need to be?

This discussion may lead into questions about the intended use of the solution. If the answer is known, you can share that information. Then have the observers report on the aspect of mathematical behavior observed, what they heard in the small-group discussions, and their analysis of what they saw happening. After each observer reports, check with the participants to see if the observations have accurately captured what was happening in their groups and ask if they have anything they would like to add.

Whole-group discussion is useful for reflecting on the experience. Questions such as the following may stimulate the discussion:

- How does observation give you data about what students know?
- How might you use observation techniques with your students?
- Which aspects of behavior would you observe?

Summary

Allow approximately five minutes. Pass out observer sheets to all the participants. Summarize the issues that were raised during the reflection period and relate the discussion explicitly to your original goals for the investigation. Proceed to the second exercise.

Second Exercise

This second exercise offers additional practice in observation skills.

Getting started

Allow approximately five minutes. Introduce the task and the context for the task. Choose two people from each table (or roughly one-third of the participants) to be observers. Be sure that the observers in this exercise worked on the task in the last exercise. Ask the observers to come to the front of the room for further instructions and distribute copies of the task to the remaining participants.

Share the observer sheet with the observers, and explain the Design Your Own Checklist sheet. The handout asks the observers to create their own observation categories. Some possibilities are *understands a particular mathematical concept; participates in discussion; responds to others' ideas or comments; uses appropriate mathematical terminology;* and *uses tools, such as a ruler or measuring tape, correctly.* The observers should choose two categories to focus on. This time, the observers will stay with one group throughout the problem, but they still should not contribute to the ongoing conversations, even when they hear misleading information or incorrect reasoning.

Experiencing and observing the task

Allow approximately forty minutes. Have the participants estimate an answer to the problem, if appropriate, and record this estimate. Then ask that the small groups create two possible strategies for solving the problem but that they not solve the problem yet. The observers should watch quietly (again, not participating) while each group decides on a solution strategy.

The small groups then record their strategies on chart paper or on an overhead projector and share them with the larger group. The observers report their observations on that discussion. Each small group then chooses one of the strategies presented (either its own or that of another group) and uses that strategy to solve the problem. Ensure that each group uses a different strategy. The observers should again watch quietly while each group solves the problem.

The groups report their findings, and the observers report their observations. The participants then reflect on whether the observations have accurately captured what was happening in their groups and add other pertinent comments.

Reflecting on the experience

Allow approximately twenty minutes. Divide this session into two parts—a presentation of solutions and a reflective discussion.

When all the small groups have decided on an answer, each group should share both the answer and the method used. Record both the answer and the method on chart paper or an overhead projector, and compare them to the group's original estimate. After all the groups have presented, ask the participants to consider which method they would choose, and why.

Next, have the observers report what they heard from the group discussions and analyze what they saw happening. Using the following or similar questions, ask several observers to comment on their experiences during this exercise:

- What challenges does conducting observations in the classroom pose for teachers?
- How else might you assess students' work on this problem?

The session should conclude with a discussion of how the participants could use this technique regularly in their classrooms. Questions to begin the discussion include the following:

- What might you learn about your students?
- How would that information guide your instruction?
- What would be the challenges to implementing this sort of technique?
- How might you address those challenges?

Summary

Allow approximately five minutes. Close both activities by summarizing the issues that were raised during the reflection period or by relating the discussion explicitly to your original goals for the investigation. Discuss possible follow-up options and next steps.

To follow up on this activity, have the participants plan to do an observation in the weeks following the workshop and reconvene to discuss their experiences and share ideas and solutions to the challenges raised.

Facilitating the Experience

If the task chosen included Fermi problems, the answers and the methods of obtaining them should vary. No one group's answer should be recognized as right or wrong.

As the observers report back in the first exercise, it may be helpful to group their observations according to the NCTM Standards (problem solving, reasoning, communication, connections, content areas) or other mathematical catagories from state frameworks. Facilitators can record this information, organized into categories, on flip-chart paper or an overhead transparency. Grouping observation data in this way will help participants see how observation techniques can be used to gather data on students' understanding of mathematical big ideas.

Instruct the observers to watch each group and make notes on what they hear that fits in their categories. Observers should not interrupt the discussion or contribute to it except perhaps to answer logistical questions. For some, this may be difficult.

The second exercise, in which observers define their categories, can be powerful in that it requires participants to identify what's important about the task—in other words, what important aspects of students' learning are represented in the task.

APPENDIX 1

Observing Groups

Make brief notes concerning one of the following points:

- Describe the mathematics the group uses.

- Describe the strategies the group uses to solve the problem.

- Comment on the group's ability to explain their methods and communicate their findings.

- Comment on the group's ability to work together to accomplish this task.

Developed by Doug Clarke, Australian Catholic University, and Linda Dager Wilson, University of Delaware. Used with permission.

Design Your Own Checklist

Choose two or three categories you will focus on as your group works on the task.

Write the group members' names in the left-hand column, and write the categories of behavior you have chosen to observe in columns 2–4. Put a mark under each category as you observe it. In the last column, add any comments you believe are useful for documenting what you observe.

Name				Comments

Developed by Doug Clarke, Australian Catholic University, and Linda Dager Wilson, University of Delaware. Used with permission.

EXAMINING STUDENTS' WORK

Overview

In traditional mathematics assessment, interpreting students' responses to problems was fairly straightforward: an answer was either correct or incorrect. Open-ended questions and other alternative assessments go further, which allows for richer responses from students and better data about students' mathematical content knowledge and problem-solving skills. This information helps to inform teachers about where students are and what instruction they need to concentrate on. In order to be useful for the teacher, those data must be skillfully interpreted. Professional experiences in collegially sorting and scoring students' work samples help teachers develop skills in interpreting evidence and using the results.

Interactions among teachers are a central part of this investigation. They work in groups to weigh the evidence in students' work samples and to make instructional or scoring decisions.

Objectives

- Teachers articulate features of quality student work and make judgments about students' work.
- Teachers make inferences about students' understanding by collegially examining evidence in students' work.
- Teachers consider what actions to take on the basis of students' work (e.g., revise the task, respond to students individually, or adjust instruction).

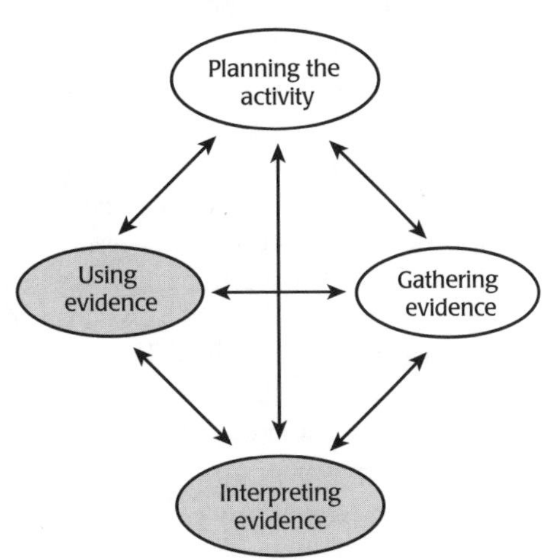

Rationale

Collegial experiences in examining students' responses to assessment items helps the individual teacher develop the skills to analyze and evaluate students' work. In the process of examining students' work together, teachers consider important questions: What responses do we expect from students to the new tasks we challenge them to accomplish? What are our standards for good mathematics work? How do we interpret students' responses? What do the data tell us about students' mathematical understandings? How can we evaluate the work? When teachers work collegially to score students' responses, they can also consider together how the assessment data should guide future instruction.

Professional discussion is an important component of the professional development activities outlined in this book. When a teacher works alone to examine and score students' work on a task, no one challenges the teacher's appraisal of students' responses. Challenges from other teachers are important because they force the evaluator to be specific about what students' misconceptions are and to articulate standards for students' work. The process opens up expectations and standards for challenge by a professional community. The opportunities for such conversations do not abound in everyday school life; most often, students' work is evaluated autonomously by the individual teacher under tight time constraints. Professional development activities can furnish those missing opportunities.

Working with groups of teachers to sort and score students' work also opens up tasks to scrutiny. If the task does not reveal clearly what students know and do not know, then it may not be worth repeating. It will surely need revision. This scrutiny creates discussion among teachers about what makes a good mathematical task for students. It can lead naturally to revising a task and trying it again in the classroom.

Designing the Investigation

The workshop description on pages 30–31 outlines how this investigation might be organized. The considerations that follow, used with the design considerations template on page 8, are intended to help you customize your own investigation.

1. *Determining the purpose.* What is your purpose for using this investigation? To create an experience in examining students' work, you must first think about why you are doing it. Your goal may be familiarizing teachers with holistic scoring and looking at students' work with an analytical framework. The tasks for which you choose to have teachers score students' responses should fit your purpose. If you are introducing open-ended questions or performance tasks or focusing on a content area such as fractions, choose tasks accordingly. Consider the underlying purpose in using the discussed assessment item. Would you like to focus on assessing to guide instruction? On monitoring students' progress? On evaluating students' achievement? Adjust your planning accordingly.
2. *Finding an appropriate task.* The selected tasks should require students to explain their reasoning or otherwise make their understandings visible in written form. A task that requires a minimal or single-number response from students gives teachers little information about their understandings and will lead to a less robust activity. The set of trial tasks in appendix 2 at the end of this investigation includes samples of tasks that have been used successfully in this activity.
3. *Obtaining samples of students' work.* Decide how you would like to obtain the students' work that you will use. One method is to have each teacher bring students' work from a predetermined task or set of tasks. This method is powerful because the teachers have the experience of trying the task with their students and then examining their own students' work. If you do decide to have the teachers bring samples of their own students' work, be sure to include in the beginning of the workshop an opportunity for them to explain their experience in using the task with their students. If the teachers bring students' work on a mixed set of tasks, then group together the teachers who used the same task.

If you choose to have the teachers bring students' work samples, we suggest that in a session prior to this one, the tasks be introduced to teachers and expectations for trying the task with students and collecting work be made clear. (See Experiencing a Task on page 14.)

It is important that the pieces of students' work to be examined during the investigation be chosen carefully. Some teachers will feel more comfortable contributing only the best responses, but encourage them to pick at random or to pick a real range of responses. This will lead to a richer discussion about students' misconceptions.

4. *Choosing a guide for analysis, such as a rubric.* An essential component of planning to use this investigation is choosing a rubric, scoring guideline, or other guide for teachers to use in examining students' work. If your district, state, or curriculum program has a holistic or analytic rubric for scoring open-ended questions, you may wish to use that rubric. If it does not, you may wish to choose a more general holistic rubric. NCSM's (National Council of Supervisors of Mathematics) publication *Great Tasks and More* (1996) offers several sample rubrics that may be useful.

   To focus participants on interpreting students' work in order to guide future instructional decisions, the two-point rubric adapted from the New Standards Project's scoring process may be helpful. The rubric appears in appendix 3 at the end of this investigation. The power of this rubric is that it leads naturally to thinking about instructional planning. You could adapt this investigation for purposes other than evaluating students' responses using a rubric. For example, students' work could be the evidence used to analyze the quality of a task and judge how well the task achieved its intended purpose. In that instance, rather than use a rubric, you could use the original purpose of the task, as described by the "presenting" or "contributing" teacher, in place of a rubric as a guide for examining the work.

5. *Choosing discussion questions.* The questions you pose are essential to focusing discussion after completing the task. The workshop description offers suggested discussion questions that focus on using the inferences made in the analysis of students' work to guide future instructional decisions. If these do not fit your purposes, you may wish to craft alternative questions.

Preparation

- We recommend that participants work in groups of four and that the room setup allow teachers to pass work easily to one another. Round or square tables are therefore preferable.
- Prepare samples of students' work. If teachers are bringing samples of their own students' work, survey the group quickly as they get settled to determine roughly the number of pieces of students' work for each possible task. This information will help you decide how to group the pieces. If you are using an already existing set of students' work, make copies of this set, one for each table or, if possible, one for each participant.
- Prepare overhead transparencies of the sorting guidelines in appendix 3 and the

tasks being sorted. If you have the work samples ahead of time, make one set of transparencies of the samples to support large-group discussion.

Materials

Photocopies of the sorting guidelines
Blank overhead transparencies
Overhead markers
Pencils (for making nonpermanent marks on students' work)
Sets of students' work

Time needed

Two and a half hours. This sorting activity can take from two to three hours. We recommend that you schedule at least two and a half hours for it and allow generous time for discussion at the end.

Workshop Description

Getting started

Allow approximately twenty minutes. Discuss the goals for the session and explain the format of the investigation. If they have not done so already, the participants should do the task and then consider what they would expect for students' responses to it. (See Experiencing a Task for suggestions on structuring this part.) If they have used the task in their own classrooms already, allow the participants to talk about how they used it with their students without yet sharing the students' responses. Have the participants arrange themselves in groups of four; if the participants have used a mix of tasks, the groups should be arranged so that those who tried the same problem with their students sit together.

Introducing the activity

Allow approximately twenty-five minutes. If the participants have brought students' work, determine the number of pieces. Each teacher should put two to four pieces of students' work into the pile to be sorted so that there are ten to fifteen samples overall. (Alternatively, if the participants are using a premade set of work, the facilitator should create one pile of responses for each group.) They should number the pieces of work (i.e., 1–15) so they can easily identify the pieces. Have the participants clear from the table all the samples that they are not using.

Hand out the sorting guidelines. Use overhead transparencies of this sheet and explain the activity. Offer a definition of a *rubric* if necessary. Answer any questions.

The investigation

Allow approximately forty-five minutes. First, teachers should evaluate and sort students' work individually, then record their ratings of each piece and any comments they have on a sheet of paper. Then the group members share their ratings with one another, discuss areas of disagreement, and work to come to a consensus.

After they have reached a consensus, the groups should discuss any misconceptions revealed in students' responses to the task. Ask each group to consider the following:

- For what content do students need sustained instruction?
- Where does their work need revision?

Each group should choose one intriguing piece of work to examine in depth. The participants should be prepared to reflect in a whole-group discussion on what that student understands and why they came to the conclusions they did about that student's work.

Reflecting on the experience

Allow approximately twenty minutes. Divide this session into two parts—a report of results and a reflective discussion.

Have each small group report to the whole group, focusing on the piece of work they examined in depth.

After each group has reported, facilitate a general large-group discussion of instructional-response strategies. You may wish to have the participants respond first to the following question in silent, individual, reflective writing before proceeding to a group conversation: How would you respond instructionally to these misconceptions?

Summary

Allow approximately fifteen minutes. Summarize the issues that were raised during the reflection period and relate the participants' comments to your original goals for the investigation. Discuss possible follow-up options and next steps.

Facilitating the Experience

If the teachers are using their own students' work, it is important that the environment be comfortable for doing so. It should be clear to the participants that you are reflecting on the students' understanding, not the adequacy of the instruction. Lay ground rules about this issue up front: make it clear to all that you recognize that this activity is risky and that you are asking them to support one another's efforts.

If a consensus on scoring is not reached, the facilitator should help characterize the areas of disagreement. When circulating among small groups or facilitating the large-group discussion, you may hear the following kinds of disagreements:

- About what students understand mathematically
- Over whether students who used an algorithm understand the underlying mathematics
- About whether the data in a student's response indicates a problem in communication or in understanding
- About overstating or understating what students know, given what they demonstrated

The facilitator can offer a group some reflection on what he or she understands a disagreement to be about. The facilitator can also encourage the group members to be specific about what evidence in the students' work leads them to make their inferences.

Some groups may finish before others. Ask early finishers to choose more pieces to read and rate. Ask them if examining more samples gives them new insights into students' thinking on the problem.

APPENDIX 2

Trial Tasks for Examining Students' Work

The open-ended tasks that follow have proved useful for teachers' initial experimentation with alternative classroom assessment. They yield good student work samples for use with this investigation.

Decimal Dilemma was developed by the Pittsburgh Public Schools and is reprinted with permission. The other tasks were developed by the Connecticut Common Core of Learning Mathematics Assessment Project, sponsored by the National Science Foundation. The tasks are reprinted with permission.

DECIMAL DILEMMA

Kiara and Jason walked home after school. They had just completed a mathematics test. It was a difficult one. Both were talking about the answers they wrote on the test. They disagreed on some of their answers.

For one of the questions, they had to choose the larger decimal: .3 or .26. Jason answered .26, but Kiara thought that .3 was larger.

Help Kiara and Jason settle their disagreement. Explain who is correct and why.

THE BUDGET MYSTERY

In 1990, the maintenance budget for a school was $30,000 out of a total budget of $500,000. In 1991, the figure was $31,200 out of a total budget of $520,000. Inflation between 1990 and 1991 was 8%.

Parents complain that the money spent on maintenance increased. The maintenance manager for the school complains that the money for maintenance decreased. The principal maintains that, in fact, there has been no change in spending patterns at the school.

Is it possible that everybody's opinion could be valid?

Write a paragraph describing how each party might justify its claim.

McDONALD'S CLAIM

You and a friend read in the newspaper that 7% of all Americans eat at McDonald's each day. Your friend says, "That's impossible!"

You know that there are approximately 250,000,000 Americans and approximately 9000 McDonald's restaurants in the U.S. You think the claim is reasonable.

Show your mathematical work and write a paragraph or two that explains your reasoning.

A WEIGHTY MATTER

The progress of three people on diets is recorded in the chart below.

Weeks on Diet	Tom	Jaime	Rhonda
0	210	158	113
2	202	154	108
4	196	150	105

Based on the trends over the first four weeks of dieting, predict each person's weight after 6 weeks on the diet. Explain your reasoning for each.

TOM ______________________

JAIME ______________________

RHONDA ______________________

Use the data in the chart to argue that:

Tom is doing the best so far because ...

Jaime is doing the best so far because ...

Rhonda is doing the best so far because ...

POSTAL RATES

Postal rates have been figured by the ounce since July 1, 1885. Since then, the rates have been:

Nov. 3, 1917	3 cents
July 1, 1919	2 cents
July 6, 1932	3 cents
Aug. 1, 1958	4 cents
Jan. 7, 1963	5 cents
Jan. 7, 1968	6 cents
May 16, 1971	8 cents
March 2, 1974	10 cents
Dec. 31, 1975	13 cents
May 29, 1978	15 cents
March 22, 1981	18 cents
Nov. 1, 1981	20 cents
Feb. 17, 1985	22 cents
April 3, 1988	25 cents
Feb. 3, 1991	29 cents

Based on the postal rates since 1917, predict the cost of mailing a one ounce first class letter in 2001. Explain your reasoning.

A HOMEWORK HELPER FOR ELENORE

You are babysitting Elenore, who has just finished her arithmetic homework. "I really like doing decimals! It's easy," the youngster boasts as she proudly shows you her work:

$$\begin{array}{r} .8 \\ + .4 \\ \hline .12 \end{array} \qquad \begin{array}{r} .3 \\ + .6 \\ \hline .9 \end{array} \qquad \begin{array}{r} .7 \\ + .9 \\ \hline .16 \end{array}$$

Is Elenore's homework correct? If not, write what you would tell and/or show Elenore, to help her *understand* her homework.

APPENDIX 3

Sorting Rubric

Ready for revision. The response accomplishes the purpose of the task or provides ample evidence that the performer has the mathematical power to do so. The execution may be faulty. An operational test for inclusion in this category is the following teacherly question: Can you give written feedback to the student that focuses his or her attention on the flaw that would be sufficient (without dialogue) for the student to revise the performance so that it would meet the standard?

More instruction needed. The response lacks adequate evidence of the learning or strategic tools that are needed to accomplish the task; or, the response has evidence of inadequate learning. Feedback to the student would not be enough. A teacher would have to interact with the student or teach more.

This rubric is used by the New Standards Project for the initial sorting of student responses to open-ended, on-demand mathematical tasks.

Sorting Students' Work

1. Each table should be working with a pile of about ten to fifteen samples of students' work.

2. Number the papers (e.g., 1, 2, 3, ..., 9, 10) so that you can easily identify each response.

3. Read through the provided rubric.

4. Each person at your table should individually record a score for each response according to the rubric.

5. After everyone has scored all the papers, discuss any disagreement on scoring. Try to come to a consensus.

6. After you have reached a consensus, discuss (as a group) the misconceptions you noted in the responses. Identify one intriguing piece to examine in depth. Be prepared to share your thinking with the entire group.

DEVELOPING TASKS

Overview

Teachers recognize that tasks—activities used for assessment purposes—have to be adapted to suit the contexts within which they teach. Such contexts may include classrooms comprising students from diverse backgrounds and with diverse needs. Students' school mathematics background is also a consideration, as is the wide range of experiences and communication skills students bring to the classroom from a variety of backgrounds. In this light, the use of ready-made tasks may not be appropriate. A more helpful mind-set among teachers is that tasks may require some revision to meet the needs of the particular students.

This investigation probes the nature of "good" assessment questions and challenges. It asks teachers to consider what they know about their students, their teaching contexts, their curricula, and their mathematical values and to align all those considerations in the development of tasks that prompt students' performance.

Objective

- Teachers establish criteria for judging the quality of tasks and develop tasks of their own.

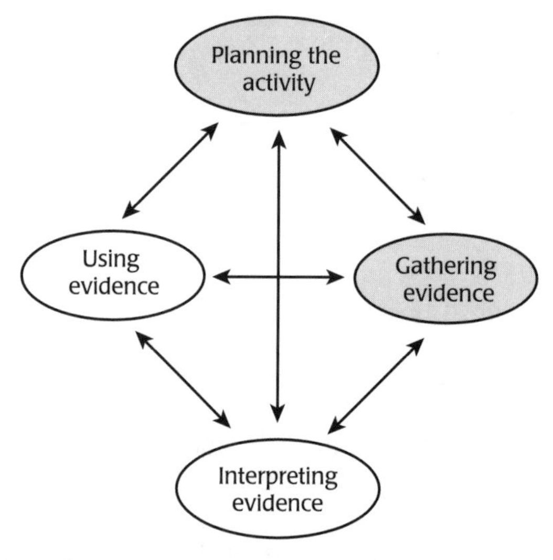

Rationale

Much of our work in professional development has focused on broadening teachers' appreciation of the purposes of assessment. In particular, we have urged increased attention to the two purposes relevant to classroom assessment: monitoring students' progress and making instructional decisions. We have found it advantageous to return time and again in discussions about assessment tasks to the question of the purpose for assigning the task. Tasks good for one purpose may not be suitable for another. Often a task or question may not be especially useful for evaluating students' achievements—for example, it may not be subject to easy scoring—but it may be very useful for uncovering students' misconceptions or for directing decisions about instruction.

Whether built around finding, revising, or creating assessment tasks, professional development focused on the quality of tasks can be beneficial learning experiences for all involved. It is important, however, to link discussions about the quality of tasks with discussions about the purposes of assessment and with analyses of students' work. After all, the final measure of whether a task is "good" depends on its suitability for the purpose, as revealed by students' performance on the task. In the end, we want to gauge tasks by studying students' work and asking questions like Was the task good for monitoring students' progress toward standards? Did the work on the task give some indicators for further instruction?

To expand teachers' repertoires of useful assessment tasks, we have employed two complementary strategies:

- Collect and share tasks for assessing mathematics learning that are already in use in various cities, states, and other countries and in curriculum projects aligned with the NCTM *Standards.*
- Build the capacity to refine and create tasks.

We do not advocate that teachers develop all their own assessment items. We do think, however, that teachers find it necessary in the course of planning for their classrooms to choose, modify, and occasionally develop tasks appropriate for their students. Experiences in task development in a professional development setting can help teachers develop their judgment about the quality and appropriateness of tasks and the performance evidence a task will provide.

A task for classroom assessment evolves as—

- a task is created or adapted;
- it is reviewed by colleagues;
- it is tried with students;
- students' responses are analyzed by the individual teacher and colleagues;
- it is revised on the basis of that analysis.

Task development is a cyclical process, with authors revising tasks on the basis of feedback from colleagues and from the evidence in students' work. An important mind-set for task development is the recognition that the process itself has value and that it takes time and requires continuous feedback.

Designing the Investigation

The workshop description on pages 38–40 is a guideline for how this investigation might be organized. The considerations that follow, used with the design considerations template on page 8, are intended to help you customize this investigation.

1. *Determining the purpose.* What is your purpose for using the investigation? Be clear about identifying your goals and desired outcomes. In particular, consider using this investigation in a sequence of inqueries. Although it may seem logical to use it early in a sequence (after all, task development seems to occur naturally early in the assessment process), we would discourage that placement. We believe that this more advanced investigation will prove more successful if participants have several opportunities beforehand to look at alternative forms of assessment and students' work.

This investigation leads quite naturally into Examining Students' Work (see p. 27). The final measure of whether a task is "good" depends on its suitability for the purpose, as revealed by students' performance on the task.

2. *Selecting materials that support the development of tasks.* If you would like this experience to connect directly with a curriculum unit the teachers are using or state or district frameworks or standards, you may want to ask teachers to bring those materials with them or make copies of the materials for them.

3. *Choosing focus questions for revising tasks.* The workshop description offers a list of questions to ask participants as they reflect on the tasks they create (see the section on writing a draft of the task). You may want to focus these questions on particular concerns. As the facilitator, you can ask the questions as you work with small groups to draft a task. The following questions may also be helpful:

   - Does the task prompt students to extend their thinking from a single answer to a general rule?
   - Does the task invite students into it at various levels of experience?
   - Is the task ambiguous? If so, is the ambiguity natural and appropriate?
   - Can solutions be produced in a straightforward manner, not just by guessing the mathematical trick?

Preparation

- We recommend that teachers work in groups of two to four.
- Identify and photocopy onto handouts, overhead transparencies, or flip charts a list of the characteristics of a good assessment task. Alternatively, have teachers create the list.
- Photocopy an outline of the task-development process.

Materials

Flip-chart paper or blank overhead transparencies

Time needed

Two and a half hours. We recommend two to three hours for this investigation; the time allocations reflect a two-and-a-half-hour workshop.

Workshop Description

Getting started

Allow approximately twenty minutes. This investigation, which incorporates ideas from Marzano, Pickering, and McTighe (1993), begins by having teachers discuss the characteristics of good tasks. Do this either by sharing a list of characteristics of good assessment and having participants add to this list or by having participants create the list. We have had teachers consider a good task that they use with their students and what makes it good. Those qualities form the list.

Next, introduce the process of developing tasks. Describe the approach the workshop will take to develop tasks that meet the listed criteria.

Identifying the task

Allow approximately fifteen minutes. Have the participants work in groups of two to four. Each pair or small group should work on developing one task. The participants should identify the following:

- The purpose of the task
- What is important for students to know and be able to do
- The processes or thinking skills involved in doing the mathematics

Encourage the participants to think about their upcoming curriculum plans so that this task relates to something they will teach in the near future. The content should connect to significant mathematical ideas and themes, such as proportional reasoning or generalizing from patterns. An example of a thinking skill required by the task might be "constructing mathematical arguments with appropriate support."

Writing a draft of the task

Allow approximately thirty minutes. The next step is using a "givens and goals" process to develop a context for the task and identify goals for students. This approach to task development, developed by Bonnie Hole as part of the Connecticut Common Core of Learning Mathematics Assessment Project, asks the participants to identify what information, context, data, or situation the task will provide ("given") and what students will be asked to do with that information ("goals"). For example:

- Given a collection of certain data, make a prediction.
- Given certain constraints, design a product.
- Given a real-world dilemma, design an experiment.
- Given an impending decision, do a survey.

Have each group write a first draft of a task. When all the groups have a draft, make sure that each writer or writing team considers the following set of germane questions:

- Does the assessment elicit the use of mathematics that is important to know and be able to do?
- Does the assessment relate to instruction and build on each student's understanding, interests, and experiences?
- Does the assessment provide modes of response that encourage each student to do the mathematics being assessed?
- How well does the problem the students will solve match the purpose of the task?
- Does the task invite multiple entry points, multiple pathways to solution, or multiple solutions?
- Is it clear to the students what they are expected to do?
- Are the students familiar with the context? Is the context or situation interesting? Is

the problem framed in a real-world context? Is the context or situation authentic or contrived?

- Is the goal accessible and appropriate for the time allowed?
- Is the audience identifiable and appropriate for the task?
- Does the task clearly invite a mathematical response?

Each group should then modify the task to include communication and information-processing skills. An example of a communication skill is "expressing ideas clearly and thoroughly," and an example of an information-processing skill is "effectively gathering, interpreting, synthesizing, and assessing the value of information."

Refining the task

Allow approximately twenty minutes. After the task has been modified, have each group exchange tasks with another group. In reviewing and working through the tasks, examine them in light of the list of characteristics of good tasks. Each group should share its impressions of the task with the authors and make suggestions for improving it.

Writing a second draft

Allow approximately twenty minutes. Using the feedback, have the groups revise their tasks and write a second draft.

Reflecting on the experience

Allow approximately twenty minutes. Gather the entire group for a reflection on developing the task. Questions such as the following may help with this discussion:

- What was difficult or challenging about developing the task?
- What changes or refinements would you still like to make to the task?
- What did you learn from the process?

The participants should commit to trying the task with students. If you plan to follow this activity with Examining Students' Work, ask the participants to collect students' work and bring it to the next workshop.

Summary

Allow five minutes. Summarize the issues that were raised and relate the discussion to the original goals of the investigation.

Facilitating the Experience

We have found it helpful to examine tasks that are good or help reveal students' thinking and to agree on a list of characteristics such as the one that follows (Sullivan and Clarke 1988, p. 14):

> "Good" questions have three features:
>
> a) They require more than recall of a fact or a reproduction of a skill.
>
> b) Pupils can learn by doing the task, and the teacher learns about the pupil from the attempt.
>
> c) There may be several acceptable answers.

Facilitators may need to help groups generate ideas for tasks. Some possible strategies for "jump starting" task development follow:

- Borrow ideas from teaching. Almost any activity that is authentic for a teacher is an authentic context for a student (e.g., checking homework, evaluating students' work, constructing a lesson).
- Have on hand sample tasks from other sources for generating ideas. (The Resources section of this book offers many sources.)
- Use reference materials, such as an almanac or other books of interesting facts or statistics, that offer data and statistics that can form the "given" in a problem.

This investigation focuses on the development of tasks. To ease into the process, teachers can start by modifying or adapting existing tasks. Abundant raw material for modification exists in prompts from traditional examinations and in the activities and exercises contained in textbooks and other instructional materials.

DEVELOPING A RUBRIC

Overview

This investigation helps teachers explore two important questions: (1) What constitutes mastery on a particular task? (2) How does a task help us discriminate among students' levels of understanding to determine if our standards for students' achievement are being met?

This investigation sharpens the discussion of what is valued by teachers as good work in mathematics. In this investigation, the teachers examine and rate students' work on a common task. Through the rating process, the teachers reach a consensus on the criteria for the levels of response.

Objectives

- Teachers define what constitutes mastery performance on a task.
- Teachers discriminate among different levels of progress toward mastery.

Planning the activity
Using evidence
Gathering evidence
Interpreting evidence

Rationale

When teachers change expectations and standards for students' mathematical work, they must carefully rethink what the criteria for "good" work are. These criteria help teachers do the following:

- Judge students' work fairly.
- Communicate expectations to students and parents.
- Most important, determine where students fall short of standards and how to address their needs instructionally.

Designing the Investigation

The workshop description included on pages 44–45 is a guideline for how this investigation might be organized. The considerations that follow, used with the design considerations template on page 8, are intended to help you customize this investigation.

1. *Identifying your purpose in using the investigation.* Be clear about identifying your intended goals and desired outcomes. Do you want to acquaint participants with

creating a criterion-referenced approach to evaluating students' work? Do you want participants to focus on identifying their criteria for performance in a particular mathematical area? Would you like participants to become familiar with a holistic rubric currently used in the school system? Are you interested in participants' adapting a general rubric for use on a particular problem? Are teachers practicing for a district- or state-level scoring process? Your purpose should guide your selection of task, rubric, and sample work.

2. *Finding an appropriate task.* As in Examining Students' Work, select a task that lends itself to producing written evidence of students' understanding and performance. Tasks that require only a single answer and do not include justifying or explaining the solution will not provide the needed evidence. The Popcorn Task, shown in figure 6 is an example of an evidence-producing task. Ideally, the participants should have had a chance to try the task you have selected in their classrooms. (See Experiencing a Task on page 14 for a suggested format.)
3. *Collecting samples students' work.* If you do not plan to have teachers bring their own students' work on a task, you may want to use work collected by your school, district, or state on an open-ended problem.
4. *Selecting a generic rubric, or scoring guide.* The basic premises of this investigation are that participants identify levels of performance generally by sorting samples of students' work and then identify their criteria for those levels. The participants first sort the students' work into high, medium, and low categories without any generic descriptors of performance, then develop the descriptors. They then sort each of the original categories into two subcategories and provide descriptors for them. The

Popcorn Task

Working with your group, estimate the number of kernels in the container on your table. You may use any of the materials available to you. The only thing you cannot do is count all the kernels in the container. Explain how you found your answer and why you think it is a good estimate.

Each group has the following materials:

An unmarked container of popcorn
A scale and weights
Several graduated cylinders of varying size

Fig. 6. A task developed by the Massachusetts Department of Education. Used with permission.

result is a six-point rubric. If you are interested in using a four-point rubric, have the group initially sort the work into two piles, and then into four. If your school, district, or state has a generic rubric, you could use it to tailor this activity for practice in adapting a generic rubric for specific use.

Preparation

- We recommend that teachers work at tables in groups of five or six.
- The primary preparation for this investigation is obtaining students' work on your task from participating teachers' classrooms or some other source. Make the appropriate number of copies (at least one set for each small group and if possible, one copy for each participant).

Materials

Blank overhead transparencies

A set of samples of students' work

Photocopies of the task and the questions used in the opening discussion

Time needed

Two to three hours

Workshop Description

Getting started

Allow approximately thirty minutes. Discuss the goals for the session, and explain the format of the investigation. If teachers have not tried the task, they should do so. See Experiencing a Task on page 14 for suggestions on structuring the activity. If the participants have tried the task in their classrooms, allow them to discuss that experience briefly, perhaps first in pairs and then in the full group.

Have the teachers discuss the following questions in small groups of five or six first and then share their ideas in a large-group discussion. Set up ten minutes of small-group discussion for each question below plus time to share the results of these discussions.

- What constitutes mastery of this task?
- How does this task discriminate among answers?

Looking at students' responses

Allow approximately forty minutes. Gather the participants into groups of five or six. Next, hand out samples of students' work on the task or have the teachers organize work they have brought. Each group should have approximately fifteen to twenty work samples.

The participants should work individually to examine all the students' responses and make individual judgments about whether each response displays high, medium, or low quality. After each participant has rated all the responses, the group should discuss the ratings and try to reach a consensus. This process will generate discussion about what is valued in the task.

Creating criteria

Allow approximately thirty minutes. In their small groups, challenge the participants to create a set of phrases that describe a high-quality response—the descriptors of proficiency, or mastery. The experience of having reached a consensus on the ratings should be beneficial in creating the descriptors. Each group should then come up with a set of descriptors for the medium- and low-quality responses. As the groups finish, ask each group to record all their descriptors on a blank transparency.

Reflecting on the experience

Allow approximately thirty minutes. Each small group should share its descriptors; all the descriptors together form a rubric. After each group has reported, facilitate a general discussion that focuses on the differences among the descriptors and moves toward a consensus on which ones to use.

Further refining the criteria

Allow approximately twenty minutes. Return the participants to small groups. Each small group should pick one performance level (high, medium, or low) to explore further. Each group should sort the responses in their level into two piles. This will create a total of six categories; if a small group is working with the original high-level responses, it should sort those papers into high-high and high-low categories. The small groups should again identify and record the criteria for their decisions on an overhead transparency.

Reflecting on the experience

Allow approximately twenty minutes. Have the groups, using their overhead transparencies, present their criteria for the six new groupings. Ask the large group if they agree with the criteria as they are presented. Try to reach a consensus. As agreement is reached in each category, record the criteria on six separate pieces of flip-chart paper hung on the wall.

Ask the group questions to test the consensus on the criteria:

- Does anyone not agree with that criterion?
- Does everyone agree that this criterion is appropriate for this level? Should it be lower? Higher?
- Were some responses difficult to rate? Do these criteria clarify the level of those performances?

Summary

Allow approximately five minutes. Summarize issues that were raised during the reflection period and relate the participants' comments to your original goals. Discuss possible follow-up options and next steps.

Facilitating the Experience

In order for teachers to develop a rubric for this task, they need a clear picture of students' performance. What is it that students should know? What evidence of students' understanding do they want to see? What does mastery on this task look like? In the first large-group discussion of these questions, record contributions on a flip chart. The chart will document the discussion and can be left hanging on the wall during the rest of the investigation as a reminder.

The participants may not all be familiar with the term *rubric*. You may want to begin by defining the concept.

When we have used this investigation with the Popcorn Task, several task-related issues have arisen: Does the teacher need to know the total number of kernels to judge the answer? If we are emphasizing the problem-solving process, is accuracy important? This investigation, since it requires the groups to reach some agreement about what is valued in students' work, raises areas of disagreement to the surface. The facilitator should be prepared to listen to these issues as they arise and open them up for comment by other participants, instead of trying to give a single answer.

One of our workshops produced the following descriptors of excellence for the Popcorn Task:

- Clearly established initial relationship between the number of kernels and volume, weight, and so on
- Had an accurate answer, within a certain tolerance
- Had a refined strategy and could explain it
- Showed appropriate use of mathematical reasoning and tools
- Used more than one strategy

PLANNING ASSESSMENT

Overview

What we value in mathematics is communicated to students in the assessments we give them. Students, for their part—and most of society, in fact—place great value on what is assessed. It is important, for the purposes of guiding instruction and monitoring students' progress, that the assessments used in the classroom yield the evidence about students' learning that would be most useful to both teachers and students.

This investigation helps participants reach clarity and consensus about valued outcomes. It also helps them align their classroom assessments with valued outcomes. In the first activity, the participants focus on a valued end-of-year or end-of-unit set of outcomes and then analyze current classroom assessments—unit tests, for example—to determine the outcomes that are valued in them and to revise them if necessary. In the second activity, the participants work backward to ensure that benchmark assessments along the way can illuminate students' progress toward the outcomes. These activities emphasize the following:

- Clearly articulated outcomes and performance criteria
- Greater clarity about assessment methods and the purposes of assessment
- Alignment of classroom assessments with desired student outcomes

Objectives

- Teachers discuss what is important for students to know and be able to do in a particular area of mathematics.
- Teachers compare this list of valued outcomes to the outcomes actually assessed by their current classroom assessments.
- Teachers identify assessments that will provide adequate and relevant evidence of students' progress toward valued outcomes.

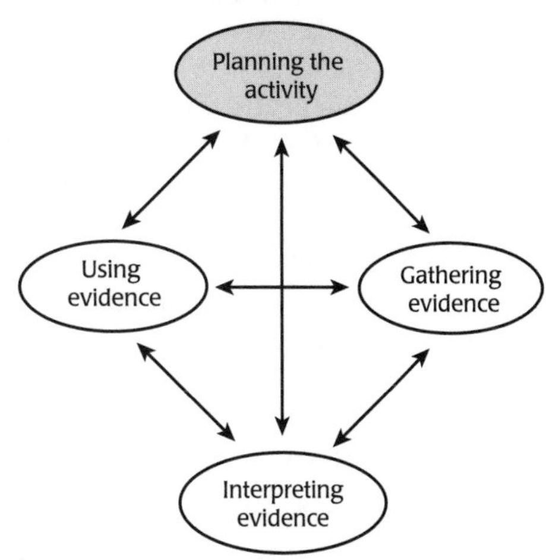

Rationale

Linking assessment with what is valuable for students to know and be able to do demands more explicit attention now that values in mathematics teaching and learning are changing. Documents such as the NCTM *Standards* and various state frameworks advocate a shift from traditional notions about the nature of curriculum, about the depth and range of school mathematics, and about the teacher's role in instruction. In light of the changes, assessment developers are hard pressed to define outcomes that match what is valued and to create assessment items that can gauge students' performances accordingly.

While assessment developers work to craft new approaches that match what is valued in mathematics, a parallel effort can and should exist in teachers' professional development. In particular, we have found it advantageous to offer formal opportunities for teachers to articulate desired learning goals for students, align their classroom assessment with those goals, and then align their instruction with their altered assessment.

Designing the Investigation

The workshop description included on pages 49–50 shows how this investigation might be organized. The considerations that follow, used with the design considerations template on page 8, are intended to help you customize your workshop.

1. *Determining the purpose.* What is your purpose for using the investigation? Be clear about identifying your goals and desired outcomes because they drive materials selection and shape after-task reflection and discussion.

   Despite our placing them in the planning phase—at the top of the diagram—the activities in this investigation shouldn't be the first step in a staff development sequence. We recommend that the participating teachers enjoy some prior experience together first, exploring the ideas and vocabulary of the assessment-reform movement.

2. *Determining the arrangement and size of the groups.* Most of the time for this activity is spent in small-group work. The small-group work is essential to the investigation; in our opinion, teachers' working together to analyze and revise an assessment plan requires them to articulate their goals and explain and test their reasoning. We suggest using pairs in this investigation; given the complexity of the content involved, we believe larger groups are potentially unwieldy, especially if all members are not working on the same unit of instruction. If you have four people who are working on the same unit, however, it may make sense for them to work together.

   Depending on the composition of the group, its members may or may not work on the same unit. This choice is up to you. If they work on the same unit, the members work together to analyze and revise assessments and to create an assessment plan. If they are not working on the same unit, they serve one another by listening, advising, and providing feedback.

3. *Arranging for appropriate materials.* The teachers should be asked to bring any materials relevant to the unit they will focus on, including any curriculum materials and assessments they have used in the past—in particular, end-of-chapter or end-of-unit tests. If you would like to target particular mathematical content or have groups of teachers address the same unit, you would need to arrange to do so before the session, perhaps in a previous session.
4. *Following up this investigation.* This investigation offers many possibilities for follow-up activities. For instance, Developing Tasks on page 36 might be an appropriate

next investigation, especially if the group needs more time to focus on aspects of task selection, modification, and development. Or, as teachers try the revised assessment plan in their classrooms, they could collect students' work for use in a future Examining Students' Work on page 27. For continuity, your intended follow-up should influence your planning.

Preparation

- Identify and photocopy appropriate pages from the NCTM *Standards* or from state frameworks.
- Ask the participants to bring with them any relevant materials, including curriculum materials and assessments associated with the unit, particularly end-of-chapter or end-of-unit tests.

Materials

Newsprint
Teachers' curriculum materials

Time needed

Two and a half to three hours

Workshop Description

Getting started

Allow approximately five minutes. Discuss the goals for the session, and explain the format of the investigation. Form the participants into groups of two. Each individual or pair should have identified in advance an instructional unit to use in the investigation. The teachers should have available to them any relevant materials: curriculum materials, assessments used in that unit in the past, and in particular, any end-of-unit tests or overall assessments of students' achievement.

Determining what to assess

Allow approximately thirty minutes. The goal is for participants to identify and articulate, in performance-related language, the intended goals for students' learning associated with the unit in question. Have the participants respond in individual, reflective writing to the question below:

- What should students know and be able to do at the end of this unit? List the important concepts and skills.

After reflecting in writing on this question, the participants should discuss their answers in pairs and modify them on the basis of each other's feedback.

They should then together make the learning goals functional by rewriting the items in performance language, using specific and descriptive verbs. For example, "use ratios" might be rewritten in performance language as "use ratios to describe relationships in tables and graphs and to develop and explain methods for solving problems." By the end of the half hour, the participants should have developed a list of goals for students' performance for the unit.

Testing the tests

Allow approximately forty-five minutes. Have each pair examine its the end-of-unit assessments and address the following questions:

- What values are communicated through these assessments?
- What desired values are not communicated?
- What important outcomes are left out?
- How might the assessments be altered to bring them more in line with the desired outcomes?

Each team should then work on modifying a question, several questions, or the overall assessment. (See the suggestions in Developing Tasks on page 36.)

Working backward

Allow approximately forty-five minutes. Have each pair work backward from the articulated learning goals to determine at what points in the unit various outcomes or learning goals will be addressed.

- At what points should the teacher have assessment information about the students' developing understandings in order to make instructional decisions?
- At what points should both teachers and students have assessment information in order to monitor students' progress toward learning goals?

Each pair should develop an assessment plan that addresses these questions. If time allows, the pairs can work to choose, develop, or revise one of these assessments.

Reflecting on the experience

Allow approximately fifteen minutes. Answering the following questions may be useful:

- What will be the most challenging part of this assessment plan to implement?
- How will you communicate the learning goals and assessment plan to students?
- Was this process useful to you?

Summary

Allow approximately five minutes. Summarize the issues that were raised during the reflection period, and relate the participants' comments to your original goals. Discuss possible follow-up options and next steps.

Facilitating the Experience

In facilitating activities like those above, some problems occasionally arise that if managed well, can become important learning opportunities:

- Phrasing performance outcomes—describing how students will show that they have learned the content we want them to learn—is difficult.
- The outcomes defined by a group of participants look, at least to the facilitator, too narrow or too shallow, or they are ones that the NCTM *Curriculum and Evaluation Standards* (1989) singles out for less attention.
- Disagreements arise between participants about what is important mathematics.

Making the learning goals or outcomes functional by using performance-oriented language can be difficult for participants. Asked "What is an important outcome?" many educators answer in the knowledge-goals parlance of traditional textbooks (e.g., "use ratios"). In order to set an assessment frame of mind among participants, facilitators need to help them change the wording of their outcomes to incorporate more specific performance language (e.g., "use ratio to describe relationships in tables and graphs, and to develop and explain methods for solving problems"). We have used such documents as the NCTM's *Curriculum and Evaluation Standards for School Mathematics* (1989) as sources for content categories. For helpful lists more specific to categories like thinking, communication, and information processing, consult *Assessing Student Outcomes* by Marzano, Pickering, and McTighe (1993).

After one of our sessions, a teacher told us, "For my twenty years of teaching, I've been letting textbook publishers determine what is important to teach. This is the first time I've been able to sit with colleagues and talk about what's important." Her reflection points to the value in making such opportunities available to teachers, but it also points to a challenge inherent in the mathematics reform movement. Communities of teachers (and administrators) must not only decide what is important but also learn how to communicate the important outcomes to one another, to students, and to the broader community of education stakeholders. The facilitator can help by asking probing questions such as "What is important for students to know in this area?" "What is important for them to be able to do?" "What kinds of performances will show that the students know and are able to do these things?" When disagreements arise, facilitators can help the participants identify and articulate the roots of the disagreements by asking clarifying questions, reflecting what they hear, and testing their understanding of the disagreement with participants.

An important emphasis for professional development is alignment—in particular, the alignment of important performance outcomes with classroom assessment but also the alignment of instruction with the desired outcomes and classroom assessment. We have found it helpful for the facilitator to persist in challenging participants to attend to how well their desired outcomes, classroom assessment, and instruction align.

Resources

Barnett, Carne. *Fractions, Decimals, Ratios and Percents: Hard to Teach and Hard to Learn?* Portsmouth, N.H.: Heinemann, 1994.

This collection of mathematics teaching cases was designed to encourage teachers to think deeply, spark inquiry about the mathematics they teach, and stimulate collaborative reflection through discussion. In a typical case, an unexpected problem or dilemma arises during class in which the students have become confused and the teacher is unsure of how to proceed. By encouraging the participants to frame problems and analyze situations, case discussions expand and deepen general knowledge about mathematics teaching and learning. That knowledge helps teachers build productively on students' thinking. The accompanying *Facilitator's Guide* offers guidelines for using the cases as a professional development tool.

Available from Heinemann, 361 Hanover Street, Portsmouth, NH 03801-3912, or call (603) 431-7874 or (800) 541-2086.

Brandt, Ronald S., ed. *Readings from "Educational Leadership": Performance Assessment.* Alexandria, Va.: Association for Supervision and Curriculum Development, 1992.

This publication is a collection of articles from recent issues of *Educational Leadership.* They address the need for change in performance assessment. They also call for consistency between educators' goals for students and how they measure students' learning. Some articles discuss the rationale for reform; some define and give examples of portfolio assessment and show how to use it. The final section offers articles on linking assessment to standards, curriculum, and desired learning outcomes.

To order, write to the Association for Supervision and Curriculum Development, 1250 North Pitt Street, Alexandria, VA 22314-1453, or call (703) 549-9110 or (800) 933-2723.

California State Department of Education. *A Question of Thinking: A First Look at Students' Performance on Open-Ended Questions in Mathematics.* Sacramento, Calif.: California State Department of Education, 1989.

This book details the results of including open-ended questions on the California Assessment Program in 1987–88. It includes a discussion of the results, a section on scoring rubrics and scoring students' responses to open-ended questions, and samples of open-ended questions and responses.

To order, call the California Department of Education Bureau of Publications, (800) 995-4099.

———. *A Sampler of Mathematics Assessment.* Sacramento, Calif.: California State Department of Education, 1994.

This publication provides information about the administration of the spring 1994 California Learning Assessment System (CLAS). It describes the process used to score the open-ended problems on the 1993 assessments as well as the four-point rubric. It also gives sample open-ended problems for each grade level tested, sample student anchor papers, and more. The mathematical problems and questions selected for the CLAS assessment integrate the mathematical strands and unifying ideas.

Copies of this publication are available from Publication Sales, California Department of Education, P.O. Box 271, Sacramento, CA 95802-0271.

Center for the Study of Testing, Evaluation, and Educational Policy. *The Influence of Testing on Teaching Math and Science in Grades 4–12.* Chestnut Hill, Mass.: Boston College, 1992.

This publication is a result of a study funded by the National Science Foundation to determine the impact of mandated testing programs on curriculum and instruction in elementary and secondary school mathematics and science education. The study took a particular interest in the impact of mandated testing on teachers having large percentages of minority students. The study analyzed widely used mandated tests in mathematics and science to determine those tests' alignment with current curricular recommendations. Other data collection included a nationwide survey of more than two thousand teachers and interviews with teachers and administrators in six urban districts. (Three other reports that resulted from the study are also available from Boston College: *The Impact of Testing on Minority Students, Teachers' and Administrators' Views of Testing Programs,* and *The Influence of Testing on Teaching Math and Science: Executive Summary.)*

To order any of these reports, call Boston College's Center for the Study of Testing, Evaluation, and Educational Policy at (617) 552-4521 or write to NSF Study, CSTEEP, 323 Campion Hall, Boston College, Chestnut Hill, MA 02167.

Charles, Randall, Frank Lester, and Phares O'Daffer. *How to Evaluate Progress in Problem Solving.* Reston, Va.: National Council of Teachers of Mathematics, 1987.

This book offers suggestions and alternatives for evaluating students' progress in problem solving. It gives explanations and examples of evaluation techniques, including observation, questioning, structured interviews, students' self-assessment through reports and inventories, holistic scoring, and multiple-choice and completion tests. Several sample strategies show how to organize and manage an evaluation program and use evaluation results.

To order, write to the National Council of Teachers of Mathematics, 1906 Association Drive, Reston, VA 20191-1593, or call (800) 235-7566.

Clarke, David. *Assessment Alternatives in Mathematics.* Canberra, Australian Capital Territory: Curriculum Development Centre, 1988.

This book outlines a professional development strategy for introducing to teachers assessment alternatives for the mathematics classroom. The author addresses many assessment techniques, both formal and informal, including annotated checklists, cumulative checklists, work portfolios, good questions, wait time, Newman error analysis, student-generated questions, investigations, students' self-assessment, the IMPACT procedure, students' journals, peer tutoring and peer assessment, group tests, student-constructed tests, and performance tasks. It also addresses communicating assessment information through grading and reports. This book is also available as a component of the Mathematics Curriculum Teaching Program assessment kit that also includes blackline masters and a volume of linked lessons.

These materials are available from Curriculum Corporation, Saint Nicholas Place, 141 Rathdowne Street, Carlton, Victoria 3053, Australia, or call 011-61-03-639-0699 or fax 011-61-03-639-1616.

Freedman, Robin Lee Harris. *Open-Ended Questioning: A Handbook for Educators.* Menlo Park, Calif.: Addison-Wesley Publishing Co., 1994.

This book takes a student-centered approach to assessment by promoting students' thinking and writing. A quick guide shows how to write open-ended questions and use different formats for questions. A guide on grading open-ended questions features sample rubrics, open-ended questions, and information on how to present open-ended questions to students.

To order, write to Addison-Wesley Publishing Company at 1 Jacob Way, Reading, MA 01867, or call (800) 552-2259.

Hart, Diane. *Authentic Assessment: A Handbook for Educators.* Menlo Park, Calif.: Addison-Wesley Publishing Co., 1994.

This book is inspired by the major revolution in assessment for all subjects. Its purpose is to help the reader understand why educators are seeking new ways to assess their students' performance and how these alternative assessments work. The alternative assessments described include portfolios, concept mapping, and performance tasks.

To order, write to Addison-Wesley Publishing Company at 1 Jacob Way, Reading, MA 01867, or call (800) 552-2259.

Harvard Mathematics Case Development Project. *Windows on Teaching: Cases of Secondary Mathematics Classrooms.* Unpublished manuscript, n.d.

These materials provide classroom-based cases for discussion and teaching notes focused on important secondary school mathematics concepts and pedagogical issues. The cases use student dialogue and written work extensively to encourage the examination of students' thinking. The teaching notes emphasize and extend the mathematical, pedagogical, and contextual issues in the cases.

For more information on the Harvard Case Development Project or these materials, contact the Harvard Project on Schooling and Children at 126 Mount Auburn Street, Cambridge, MA 02138, or call (617) 496-6883.

Herman, Joan L., Pamela R. Aschbacher, and Lynn Winters. *A Practical Guide to Alternative Assessment.* Alexandria, Va.: Association for Supervision and Curriculum Development, 1992.

This publication offers some practical suggestions and resources for districts or departments working to revise assessment methods. The guide features information and suggestions on creating assessment tasks, including determining the purposes for tasks, selecting tasks, setting scoring criteria, ensuring reliable scoring, and using alternative assessment for decision making. The information speaks to assessment broadly and is not mathematics-specific. It focuses on creating valid and authentic assessments for evaluating students.

To order this publication, write to the Association for Supervision and Curriculum Development at 1250 North Pitt Street, Alexandria, VA 22314-1453, or call (703) 549-9110 or (800) 933-2723.

Hibbard, K. Michael, and colleagues. *A Teacher's Guide to Performance-Based Learning and Assessment.* Alexandria, Va.: Association for Supervision and Curriculum Development, 1996.

This guide was written by teachers for teachers in all grade levels. It features more than forty assessment tasks—portfolios, exhibitions, and writing assignments. Tools for grading and reporting, such as rubrics and assessment lists, are also included. The guide offers in-depth explanations of ways to design performance tasks and use these assessment tools, including data on their impact on students and parents.

Available from Association for Supervision and Curriculum Development, 1250 North Pitt Street, Alexandria, VA 22314-1453, or call (703) 549-9110 or (800) 933-2723.

Lovitt, Charles, and Doug Clarke. *Mathematics Curriculum Teaching Program Professional Development Package: Activity Bank, Volume 1* and *Volume 2.* Canberra, Australian Capital Territory: Curriculum Development Centre, 1988.

Volume 1 is a collection of suggested activities and professional development structures to help mathematics educators implement reform by showing what high-quality learning environments look like. Topics include social issues, physical involvement, students' writing, mental computation, emphasizing concept learning, and the role of video.

Volume 2 includes such topics as visual imagery, estimation, story-shell frameworks, group investigation and problem solving, mathematical modeling, the use of computers, and iteration through numerical methods.

Both volumes are available from Curriculum Corporation, Saint Nicholas Place, 141 Rathdowne Street, Carlton, Victoria 3053, Australia, or call 011-61-03-639-0699 or fax 011-61-03-639-1616.

Madaus, George F., and Ann G. A. Tan. "The Growth of Assessment." In *Challenges and Achievements of American Education,* 1993 Yearbook of the Association for Supervision and Curriculum Development, edited by Gordon Cawelti, pp. 53–79. Alexandria, Va.: Association for Supervision and Curriculum Development, 1993.

This chapter, part of the 1993 ASCD yearbook looking at the past fifty years in American education, takes a historical look at the growth of testing and its role in education. The chapter covers topics such as the technology of testing before and since 1943, beliefs about learning, growth in state-mandated testing and test sales, changes in the use of tests, and legislation regarding testing.

To order, write to the Association for Supervision and Curriculum Development, 1250 North Pitt Street, Alexandria, VA 22314-1453, or call (703) 549-9110 or (800) 933-2723.

Marzano, Robert J., Debra Pickering, and Jay McTighe. *Assessing Student Outcomes: Performance Assessment Using the Dimensions of Learning Model.* Alexandria, Va.: Association for Supervision and Curriculum Development, 1993.

Performance assessment and learning-centered education come together in this book. It describes five dimensions of learning and how they work together, particularly how they relate to assessment. The chapters discuss the changing face of educational assessment, assessment standards linked to the five dimensions of learning, how performance is assessed, and using rubrics in performance assessment.

To order, write to the Association for Supervision and Curriculum Development, 1250 North Pitt Street, Alexandria, VA 22314-1453, or call (703) 549-9110 or (800) 933-2723.

Mathematical Sciences Education Board (MSEB). *For Good Measure: Principles and Goals for Mathematics Assessment.* Washington, D.C.: Mathematical Sciences Education Board, 1991.

This document resulted from an MSEB-sponsored National Summit on Mathematics Assessment in spring 1991. The principles and goals for mathematics assessment outlined here are the result of the discussions and consensus-reaching process that began at that meeting.

To order, write to the Mathematical Sciences Education Board, Harris 476, 2101 Constitution Avenue, NW, Washington, DC 20418, or call (202) 334-3294.

———. *Measuring Up: Prototypes for Mathematics Assessment.* Washington, D.C.: National Academy Press, 1993.

This MSEB publication provides prototype tasks and rubrics that give some ideas for what authentic assessment could look like across the mathematics curriculum. This book offers background information on authentic assessment and the use of rubrics. It also gives details about thirteen prototype assessments for fourth grade, including the task itself, a rationale for the use of the task, a commentary on the design considerations involved, and a discussion of scoring the task, including samples of students' work. The discussion of design parameters that follows the presentation of each task may be of particular interest to task developers.

Measuring Up can be ordered through the National Academy Press, 2101 Constitution Avenue, NW, Box 285, Washington, DC 20005, or call (800) 624-6242 or (202) 334-3313.

———. *Measuring What Counts: A Conceptual Guide for Mathematics Assessment.* Washington, D.C.: National Academy Press, 1993.

This publication adds to the national discussion on mathematics assessment, establishes research-based connections between standards and assessment, and discusses three primary principles for assessment programs: the content principle, the learning principle, and the equity principle. The message is clear: "Assessment in support of standards must not only measure results, but also must contribute to the educational process itself."

Measuring What Counts can be ordered through the National Academy Press, 2101 Constitution Avenue, NW, Box 285, Washington, DC 20005, or call (800) 624-6242 or (202) 334-3313.

Miller, Barbara, and Ilene Kantrov. *Facilitating Cases in Education.* Portsmouth, N.H.: Heinemann, forthcoming a.

This guide for facilitators of case discussions provides general support for planning and leading professional development experiences for educators centered on the discussion of cases. It offers both an overall approach to the use of cases and practical suggestions.

For more information, contact Heinemann, 361 Hanover Street, Portsmouth, NH 03801-3912, or call (603) 431-7874 or (800) 541-2086.

———, eds. *Cases on School Reform.* Portsmouth, N.H.: Heinemann, forthcoming b.

This collection of six cases on school reform covers a broad range of topics and issues. Several of the cases focus on mathematics reform issues, and some of those particularly address changes in mathematics assessment practices. Each case includes a facilitator's guide intended to help the facilitator plan and lead case discussions.

For more information, contact Heinemann, 361 Hanover Street, Portsmouth, NH 03801-3912, or call (603) 431-7874 or (800) 541-2086.

Moon, Jean, and Linda Schulman. *Finding the Connections.* Portsmouth, N.H.: Heinemann, 1995.

Designed to provide a comprehensive background to both theoretical and practical aspects of assessment, this book emphasizes the process-oriented nature of alternative assessment. The topics address the many connections found among assessment, curriculum, instruction, students, parents, teachers, schools, and communities. Supported by examples of students' work, the book addresses how students learn, creating open-ended problems, and evaluating progress.

Available from Heinemann, 361 Hanover Street, Portsmouth, NH 03801-3912, or call (603) 431-7874 or (800) 541-2086.

Mumme, Judy. *Portfolio Assessment in Mathematics.* Santa Barbara, Calif.: Regents, University of California, 1990.

This booklet introduces and explores using portfolio assessment in mathematics. Offering many examples of students' work, it addresses the purpose and content of portfolios, the selection of items for inclusion in portfolios, making sense of a portfolio, translating portfolios into grades, and the benefits and challenges of portfolio use.

To obtain this document, write to the California Mathematics Project, University of California, 522 University Road, Santa Barbara, CA 93106.

National Council of Supervisors of Mathematics. *Great Tasks and More! A Source of Camera-Ready Resources on Mathematics Assessment.* Golden, Colo.: National Council of Supervisors of Mathematics, 1996.

This book includes camera-ready resources on a variety of mathematics assessment topics: collections of released tasks for elementary, middle, and high school; sample performance-based final examinations; blackline transparency masters on making the case for reform in mathematics assessment; and others.

To order, write to National Council of Supervisors of Mathematics, P.O. Box 10667, Golden, CO 80401.

National Council of Teachers of Mathematics. *Assessment in the Mathematics Classroom.* 1993 Yearbook of the National Council of Teachers of Mathematics, edited by Norman L. Webb. Reston, Va.: National Council of Teachers of Mathematics, 1993.

This 1993 NCTM yearbook is a collection of articles on various aspects of assessment. It includes articles on classroom assessment in general, assessment at all K–12 grade levels, and issues in classroom assessment. The articles include many examples of assessment items.

To order, write to the National Council of Teachers of Mathematics, 1906 Association Drive, Reston, VA 20191-1593, or call (800) 235-7566.

———. *Assessment Standards for School Mathematics.* Reston, Va.: National Council of Teachers of Mathematics, 1995.

This publication presents six assessment standards (mathematics, learning, equity, openness, coherence, and inferences) and four purposes for assessment (monitoring students' progress, making instructional decisions, evaluating students' achievement, and evaluating programs). It includes examples, cases, students' work, and questions to guide the implementation of each standard.

To order, write to the National Council of Teachers of Mathematics, 1906 Association Drive, Reston, VA 20191-1593, or call (800) 235-7566.

———. *Curriculum and Evaluation Standards for School Mathematics.* Reston, Va.: National Council of Teachers of Mathematics, 1989.

The *Curriculum and Evaluation Standards* was developed by the National Council of Teachers of Mathematics to help improve the quality of school mathematics by establishing a vision of what content priority and emphasis a mathematics curriculum should include. The document details forty curriculum standards that make recommendations at the K–4, 5–8, and 9–12 levels. The document also lists fourteen standards for the evaluation of students and programs.

To order, write to the National Council of Teachers of Mathematics, 1906 Association Drive, Reston, VA 20191-1593, or call (800) 235-7566.

———. *Professional Standards for Teaching Mathematics.* Reston, Va.: National Council of Teachers of Mathematics, 1991.

Designed to provide, with the *Curriculum and Evaluation Standards,* a broad framework to guide school mathematics reform, the *Professional Teaching Standards* presents a vision of what teaching should entail to support the changes outlined in the *Curriculum and Evaluation Standards.* The Standards offer recommendations on choosing or creating worthwhile mathematical tasks, creating and sustaining classroom discourse, developing a supportive classroom environment, and assessing students' learning to guide instructional decision making.

To order, write to the National Council of Teachers of Mathematics, 1906 Association Drive, Reston, VA 20191-1593, or call (800) 235-7566.

———. *Standards Blackline Masters and Presentation Guide.* Reston, Va.: National Council of Teachers of Mathematics, 1996.

This collection of blackline masters for overhead transparencies and guide to presentations was created to help disseminate the NCTM *Curriculum and Evaluation Standards, Assessment Standards for School Mathematics,* and *Professional Teaching Standards.* The set includes an overview of the Standards and a potential sequence and structure for using the transparencies in a presentation.

To order, write to the National Council of Teachers of Mathematics, 1906 Association Drive, Reston, VA 20191-1593, or call (800) 235-7566.

Pandey, Tej. *A Sampler of Mathematics Assessment.* Sacramento, Calif.: California Department of Education, 1991.

The booklet, prepared along with the mathematics components of the former California Assessment Program, describes methods of assessment intended to foster and mirror good instruction. The document describes and illustrates four types of assessment (open-ended problems, enhanced multiple-choice questions, investigations, and portfolios) and includes a set of sample problems and a list of annotated references.

To order, contact the publications department of the California Department of Education in Sacramento, California, at (916) 445-1260.

Regional Educational Laboratory Network Program on Science and Mathematics Alternative Assessment. *A Toolkit for Professional Developers: Alternative Assessment.* Portland, Oreg.: Northwest Regional Educational Laboratory, 1994.

The *Toolkit* offers resources for those designing and implementing professional development experiences that support investigating and using alternative assessment. The kit includes all the materials needed for the professional development activities, from background information to copies of appropriate overhead transparencies.

The *Toolkit* is available from the Northwest Regional Educational Laboratory, 101 Southwest Main Street, Suite 500, Portland, OR 97204, or phone (503) 275-9500.

Stenmark, Jean Kerr. *Assessment Alternatives in Mathematics: An Overview of Techniques that Promote Learning.* Berkeley, Calif.: EQUALS, Lawrence Hall of Science, 1989.

This book investigates mathematics assessment alternatives, including students' mathematical products (portfolios, writing in mathematics, investigations, sample assessment items, open-ended questions), performance assessment (observations, interviews, asking questions), and students' self-assessment. It gives samples of assessment techniques, discusses issues in assessment, and suggests strategies for change.

To order, write to EQUALS, Lawrence Hall of Science, University of California at Berkeley, Berkeley, CA 94720, Attn.: Assessment Booklet, or call (510) 642-1823.

———, ed. *Mathematics Assessment: Myths, Models, Good Questions, and Practical Suggestions.* Reston, Va.: National Council of Teachers of Mathematics, 1991.

This booklet contains examples of assessment techniques that focus on students' thinking: performance assessment, observations, interviews, conferences, questions, and mathematics portfolios. Particular attention is paid to implementing these models in mathematics classrooms.

To order, write to the National Council of Teachers of Mathematics, 1906 Association Drive, Reston, VA 20191-1593, or call (800) 235-7566.

Stiggins, Richard, and Nancy Faires Conklin. *In Teachers' Hands: Investigating the Practices of Classroom Assessment.* Albany, N.Y.: State University of New York, 1992.

This book describes a program of research conducted at the Northwest Regional Educational Laboratory to investigate the complexities of classroom-based assessment. This research seeks to balance the attention to large-scale, national, nonclassroom assessments with an increased understanding and attention to the majority of assessments that affect student outcomes: teacher-developed, classroom-based assessment.

To order, write to State University of New York Press, c/o CUP Services, P.O. Box 6525, Ithaca, NY 14851, or call (607) 277-2211.

Sullivan, Peter, and David Clarke. "Asking Better Questions." *South East Asian Journal of Science and Mathematics Education* 11 (June 1988): 14–18.

This article characterizes "good" questions: they require more than recall, pupils can learn from doing the task, and there are several acceptable answers. It offers two methods for constructing good questions that involve modifying traditional approaches to asking questions.

Tsuruda, Gary. *Putting It Together: Middle School Math in Transition.* Portsmouth, N.H.: Heinemann, 1994.

Written by a California middle school teacher, the book reveals the author's professional struggles in making a transition in his teaching methods. He details the paradigm shift in his own thinking about mathematics teaching, and he describes the approaches he now takes to teaching problem solving. The book contains many activities from his classroom such as sample problems of the day, problems of the week , writing prompts, and portfolio guidelines.

To order, write to Heinemann, 361 Hanover Street, Portsmouth, NH 03801-3912, or call (800) 541-2086.